BEHAVIOUR, DEVELOPMENT AND EVOLUTION

Behaviour, Development and Evolution

Patrick Bateson

https://www.openbookpublishers.com

© 2017 Patrick Bateson

The text of this book is licensed under a Creative Commons Attribution 4.0 International license (CC BY 4.0). This license allows you to share, copy, distribute and transmit the text; to adapt it and to make commercial use of it providing that attribution is made to the author (but not in any way that suggests that he endorses you or your use of the work). Attribution should include the following information:

Bateson, Patrick, *Behaviour, Development and Evolution*. Cambridge, UK: Open Book Publishers, 2017. http://dx.doi.org/10.11647/OBP.0097

In order to access detailed and updated information on the license, please visit https://www.openbookpublishers.com/product/490#copyright

Further details about CC BY licenses are available at https://creativecommons.org/licenses/by/4.0

All external links were active on 1/1/2017 unless otherwise stated and have been archived via the Internet Archive Wayback Machine at https://archive.org/web

Updated digital material and resources associated with this volume are available at https://www.openbookpublishers.com/product/490#resources

Every effort has been made to identify and contact copyright holders and any omission or error will be corrected if notification is made to the publisher.

ISBN Paperback: 978-1-78374-248-6
ISBN Hardback: 978-1-78374-249-3
ISBN Digital (PDF): 978-1-78374-250-9
ISBN Digital ebook (epub): 978-1-78374-251-6
ISBN Digital ebook (mobi): 978-1-78374-252-3
DOI: 10.11647/OBP.0097

Cover image: Johannes Itten, 'Cerchi' (1916) © Galleria Nazionale d'Arte Moderna e Contemporanea, Rome, all rights reserved. Su concessione del Ministero dei Beni e delle Attività Culturali e Ambientali e del Turismo. Photo by Anna Gatti.

All paper used by Open Book Publishers is SFI (Sustainable Forestry Initiative), and PEFC (Programme for the endorsement of Forest Certification Schemes) Certified.

Printed in the United Kingdom, United States and Australia
by Lightning Source for Open Book Publishers (Cambridge, UK).

Contents

Preface ... 1

1. Appearance of Design ... 9
   Design of machines ... 12
   Conflicts in motivation ... 15
   Conclusions ... 17

2. Imprinting and Attachment ... 19
   Attachment in humans ... 21
   Imprinting in the wild ... 22
   Individual recognition ... 24
   [handwritten: & Conclusions]

3. Rules and Reciprocity ... 27
   Models of development ... 28
   Alternative pathways ... 31
   Rules for changing the rules ... 33
   Coordination in development ... 35
   Conclusions ... 36

4. Discontinuities in Development ... 39
   Loss of continuity ... 43
   Conclusions ... 45

5. Early Experience and Later Behaviour ... 47
   Washing the brain ... 50
   Neurobiology ... 53
   Continuity and change ... 54
   Conclusions ... 55

6. Communication between Parents and Offspring ... 57
   Parents and offspring ... 61
   Conclusions ... 66

7. Avoiding Inbreeding and Incest	67
Early experience and sexual attraction	72
Incest taboos	73
8. Genes in Development and Evolution	77
Genes in development	79
Heritability	82
Epigenetics	84
Selfish genes	87
Conclusions	89
9. Active Role of Behaviour	91
Environmental change	97
Conclusions	99
10. Adaptability in Evolution	101
Behaviour and evolution	106
Conclusions	113
11. Concluding Remarks	115
Index	121

[handwritten annotation near Incest taboos: "� conclusion"]

Preface

The effectiveness of education, the role of parents in shaping the characters of their children, the causes of violence and crime, and the roots of personal unhappiness are matters that are central to humanity. Like so many other fundamental issues about human existence, they all relate to behavioural development. The catalogue continues. Do bad experiences in early life have a lasting effect? Is intelligence inherited? Can adults change their attitudes and behaviour? When faced with such questions, many people want simple answers. They want to know what really makes the difference.

Development presents many wonders, but one of the most remarkable is how a fully functional individual grows from a microscopic embryo. The processes that are involved have often seemed beyond understanding and, even now, much remains to be discovered. Nevertheless, the factual certainties of stability and change have been known for a long time. The robust constancies of development are profound and real. Nobody will confuse a human with a rhesus monkey. At the same time, the plasticity of each individual is as remarkable as his or her robustness. Humans possess great capacity for change, a capacity that, as in other species, emerges very early in development. It does not follow, though, that two distinct processes can be cleanly separated, one leading to invariant outcomes and the other generating differences between individuals due to culture, education and experience. If such separation were possible, it might be sensible to ask the question how much of a behaviour pattern is innate and how much is learned or, more generally, how much is genetic and how much is due to the environment. This dichotomy, which was popular in the early days of my own subject

© 2017 Patrick Bateson, CC BY 4.0   http://dx.doi.org/10.11647/OBP.0097.12

of ethology, is neither valid nor helpful but unfortunately it persists in popular accounts of where behaviour comes from and in some scientific literature. I suspect that some of the persistence is due to a hangover from folk psychology and folk biology. I also suspect that some cultural lag has occurred partly because dichotomies are easy to remember and understand.

In ethology, many of us were bird watchers before we started our careers as scientists and were accustomed to the free-flowing run of behaviour of animals. Although we wouldn't have thought of it in these terms, we were accustomed to what we now see as systems.[1] The actions of one moment become the triggers of the next and feed back so that behaviour brings to an end its own performance. Other scientists have been trained analytically and assume that any research programmes should hunt down the crucial factor that produced a qualitatively distinct effect. The talk of systems may sound to them like so much waffle. Their mantra is that science is about uncovering causes.

Changing minds is always difficult, but it is possible to be optimistic that systems approaches will become widespread. Indeed, in recent years the mood has started to change. Experimentalists are less likely these days to hold all but one variable constant and, when a single independent variable is found to produce an effect, it is not immediately taken to be *the* cause, nor is everything else deemed unimportant. The nature of the feedback in free-running systems is such that the experimentalist's distinction between independence and dependence evaporates. The dependent variable of a moment ago becomes the independent variable of the present.

Maybe these changes in thinking have come about because computer literacy has made it possible to think about the interplay between many different things with comparative ease. It is not difficult to construct simple working models on our personal computers. When the rules of operation are non-linear, the behaviour of these models, when the parameters are altered, can change in complicated ways that are difficult to predict. Without basing them rigorously on what is known about behaviour and underlying mechanisms, such models merely serve to teach us a simple lesson about causality. But the more general

---

1 See Oyama, S., Griffiths, P.E. & Gray, R.D. (eds.) (2001), *Cycles of Contingency: Developmental Systems and Evolution*. Cambridge, MA: MIT Press.

point is that the development of individuals is readily perceived as an interplay between them and their environments. The current state of the individual influences which genes are expressed, and also impacts on the social and physical world. Individuals are then seen as choosing and changing the conditions to which they are exposed.

A central theme in biology has been the ways in which the various features of an organism all fit together to create a well adapted whole. Charles Darwin's great theory of natural selection provided a cogent way of thinking about how such adaptations evolved. Organisms are highly adaptable and their abilities to meet environmental challenges are also represented in the fit between their characteristics and their ecology. The appearance of design strikes us again and again and is the basis for the first chapter, but the theme runs throughout the essays in this book.

My own research interest, starting as a graduate student, has been in the development of behaviour, with a particular focus on the remarkable process of imprinting in birds. Part of this was a long-standing collaboration with Gabriel Horn on the neural basis of imprinting[2] but my work also had a strong whole-animal dimension to it. I was trained as a zoologist and frequently ask questions about the biological function and evolution of behaviour. In the second chapter I describe imprinting as a system well adapted by evolution to its current use and central to the attachment of offspring to one or both of their parents.

Chapter 3 deals with the rules that underlie the development of the individual and the reciprocity between those rules and the individual's experience. I was much influenced by the writings of C.H. Waddington[3] whose systems approach was not fashionable in the last decades of the twentieth century but now becomes increasingly important in making sense of the complexity of development.[4]

The young organism has to deal with the challenges that meet it as it develops. Its ecology may be very different from that of the adult, in which case it may have special adaptations to deal with those conditions.

---

2   Bateson, P. (2014), Thirty years of collaboration with Gabriel Horn. *Neuroscience and Biobehavioral Reviews* 50, 4–11, http://dx.doi.org/10.1016/j.neubiorev.2014.09.019
3   Waddington, C.H. (1957), *The Strategy of the Genes*. London: Allen & Unwin.
4   Capra, F. & Luisi, P.L. (2014), *The Systems View of Life*. Cambridge: Cambridge University Press, https://doi.org/10.1017/cbo9780511895555

Like a caterpillar metamorphosing into a butterfly, even a human child has adaptations to deal with each stage of its life cycle. Sometimes the changes from one stage to the next are marked by discontinuities. These are the subject of Chapter 4.

Despite the changes in the individual's repertoire of behaviour as it grows up, early experience can have long lasting effects on its preferences and habits when it finally matures. These aspects of its behaviour are often very stable, but in stressful conditions they may change when the stress is accompanied by new forms of experience. The change is usually adaptive to cope with a world that may be very different from the one in which the individual grew up. The phenomenon is explored in Chapter 5.

In mammals, parent and offspring are often thought to be in conflict. The communication between them takes the form of mutual manipulation. The offspring seeks to gain maximum advantage from its parent, and the parent seeks to defend its long-term reproductive interests if it is able to have more than one offspring. This argument is explored in Chapter 6 after a brief review of the nature of communication in animals. The conclusion is that parents do well to take into account the condition of their offspring and the offspring must likewise pay attention to the condition of their parent.

Many animals choose their mates carefully. This is especially true in birds and many mammals. Inbreeding has costs but so too does outbreeding too much. The way in which an optimal balance is achieved is in part by the experience of close kin in early life. The role of imprinting-like processes is described in Chapter 7. Is avoidance of inbreeding the same as the avoidance of incest found in most human societies? I argue that it is not. The taboos may be an expression of conformism directed at individuals doing what most people would not do.

The enormous success of molecular biology has led to the prominence given to the role of genes in development. Genes in their different guises are unquestionably needed for the inheritance of much behaviour. I argue in Chapter 8 that the importance of genes does not mean that a simple link can be found between genes and behaviour. Unfortunate metaphors such as genes providing a blueprint for behaviour have proved extremely misleading. I return to the theme, first outlined in Chapter 3, that understanding development requires a systems approach which takes into account all the genes and environmental inputs that affect development.

Organisms do not simply react to changes in the environment. They play an active role in choosing and controlling the optimal conditions for themselves. By their activities in early life they prepare themselves for becoming an adult. An important aspect of such behaviour is play. This is a subject that formed another part of my own research life.[5] These aspects of behaviour are discussed in Chapter 9 and provide a bridge to the next chapter.

Organisms' adaptability provides a major part of the link between development and evolution. This link is the subject of Chapter 10. Of central importance is understanding the relationship between what an individual does and how its activities might influence the genomes of its descendants. This issue is still a relatively under-researched area because development and evolution have usually been thought to be separate topics.

In the final chapter I pull the threads together. Inevitably many aspects of behavioural development are omitted.[6] My book presents an approach that is deeply embedded in ethology[7] as I attempt to bring together many of the factors that affect the development of behaviour. I then relate the results to their function and their role in biological evolution. The changes in thinking have important implications for the relations between the biological and social sciences.

My original intention in planning this book was to republish essays that had first appeared in multi-authored books. I am not alone in rarely reading such chapters written by others so I can hardly complain if others do not cite my chapters. As I started work, I felt the need to rewrite the essays in order to make a more cohesive book. I also wanted the ideas to be available to a wide group of people who are interested in where our behaviour comes from and what effects behaviour has on

---

[5] Bateson, P. (2015), Playfulness and creativity. *Current Biology*, 25:1, R12-R16, http://dx.doi.org/10.1016/j.cub.2014.09.009

[6] Many accessible essays about behavioural development are given in Blumberg, M.S., Spencer, J.P. & Shenk, D. (eds.) (2016), *How We Develop: Developmental Systems and the Emergence of Behavior*. WIREs Cognitive Science.

[7] Bateson, P. (2015), Human evolution and development: an ethological perspective. In: Overton, W.F. & Molenaar, P.C.M. (eds.), *Handbook of Child Psychology and Developmental Science. Vol.1 Theory and Method*. 7th ed. Hoboken, NJ: Wiley, pp. 208–243.

evolution. I stripped out much of the scholarly apparatus such as long lists in the text of authors whose work I had depended upon. I also disregarded potential accusations of self-plagiarism since much of the material would not otherwise have been available to the public.

Inevitably I am indebted to a large number of friends and colleagues for their influence on my thought. I want to mention here some people for whom I have special affection and to whom I am especially indebted. These are Niko Tinbergen,[8] Robert Hinde[9] and Gabriel Horn.[10] In different ways they provided the inspiration for a major part of my own research.

I have co-authored three books with Paul Martin[11] who was originally my student and, for some years afterwards, my colleague. After leaving

---

8   In my first year at Cambridge I went to the Edward Grey Institute student conference held each year in Oxford. Chris Plowright, who was also at the conference, and I hatched a plan to take an expedition to Spitsbergen to study the rare Ivory Gull. Niko Tinbergen, who was engaged in a comparative study of gull behaviour was keen to join us because Ivory Gulls often nest on cliffs and might have special cliff-nesting adaptations. Niko spent considerable time with us, planning what we should do. To our sorrow and his, he was prevented by illness from joining the expedition. When we returned, he gave us much help as we wrote up our results for publication. After that experience, I was set on doing research for a doctorate with him at Oxford and in my final year as an undergraduate spent some time at his field site. In the end, however, I stayed in Cambridge to do my post-graduate research. But Niko's interest in the biological function of behaviour remained with me thereafter.

9   Robert Hinde supervised my postgraduate research on behavioural imprinting. He was a superb supervisor, taking tremendous trouble over the written work of his research students. He taught us how to think. Robert exerted an extraordinary influence on ethology, primatology and latterly on studies of human behavioural biology and development. He wanted his research to be of use to humanity and had a deep concern about the causes of aggression and the peculiarly human institution of war. He was a wonderful friend and colleague throughout my career.

10 Gabriel Horn had a long-standing interest in the brain going back to his student days at Birmingham where he had written a brilliant essay on the neurological basis of thought. He had been working on attention and habituation but was very interested in the effects of learning on the nervous system. I met him for the first time at a dinner in our Cambridge college where we were both Fellows. As Gabriel and I talked animatedly, we realised that imprinting in naïve chicks would be an excellent form of learning in which to study the neural basis of memory. We agreed to work together. Thus started a warm and lasting friendship and scientific collaboration that continued for the next thirty years.

11 Bateson, P. & Martin, P. (1999), *Design for a Life: How Behaviour Develops*. London: Jonathan Cape. Martin, P. & Bateson. P. (2007), *Measuring Behaviour: An Introductory Guide*. 3rd ed. Cambridge: Cambridge University Press, http://dx.doi.org/10.1017/cbo9780511810893. Bateson, P. & Martin, P. (2013), *Play, Playfulness, Creativity*

academia, he continued to publicise many different aspects of science in outstanding surveys.[12] The books he wrote with me were genuinely collaborative and his stimulation and good sense were invaluable in our joint projects. I am deeply grateful to him, all the more so because he commented on a complete draft of this book. Much of what we wrote about both together and separately pertains to human existence. These topics are taken up in many of the chapters in this book. Another dear friend Michael Yudkin also read critically the whole draft of the book, not once but twice, and my gratitude to him is profound.

Patrick Bateson
Cambridge, October 2016

---

     and Innovation. Cambridge: Cambridge University Press, https://doi.org/10.1017/cbo9781139057691

12   Martin, P. (1997), *The Sickening Mind: Brain, Behaviour, Immunity and Disease*. London: HarperCollins. Martin, P. (2002), *Counting Sheep: The Science and Pleasures of Sleep and Dreams*. London: HarperCollins. Martin, P. (2005), *Making Happy People: The Nature of Happiness and its Origins in Childhood*. London: Fourth Estate. Martin, P. (2008), *Sex, Drugs & Chocolate: The Science of Pleasure*. London: Fourth Estate.

1. Appearance of Design

In everyday life design implies a beneficial means to an end. The idea of design has been central to much discussion in biology. Bishop William Paley in the early nineteenth century wrote about the reactions of a person discovering a watch on a mountainside, pondering on how it was made.[1] Paley wrote: 'It is the suitableness of these parts to one another; first, in the succession and order in which they act; and, secondly, with a view to the effect finally produced'. Paley emphasized how different parts of an animal's body relate to each other and contribute to the whole.[2] He regarded the design he saw everywhere in nature as proof of the existence

---

1   Paley, W. (1802), *Natural Theology*. London: Faulder.
2   Paley illustrated his idea of relations by considering the various features of the mole: 'The strong short legs of that animal, the palmated feet armed with sharp nails, the pig-like nose, the teeth, the velvet coat, the small external ear, the sagacious smell, the sunk protected eye, all conduce to the utilities or to the safety of its underground life. … The mole did not want to look about it; nor would a large advanced eye have been easily defended from the annoyance to which the life of the animal must constantly expose it. How indeed was the mole, working its way under ground, to guard its eyes at all? In order to meet this difficulty, the eyes are made scarcely larger than the head of a corking-pin; and these minute globules are sunk so deeply in the skull, and lie so sheltered within the velvet of its covering, as that any contraction of what may be called the eye-brows, not only closes up the apertures which lead to the eyes, but presents a cushion, as it were, to any sharp or protruding substance which might push against them. This aperture, even in its ordinary state, is like a pin-hole in a piece of velvet, scarcely pervious to loose paricles of earth. Observe then, in this structure, that which we call relation. There is no natural connection between a small sunk eye and shovel palmated foot. Palmated feet might have been joined with goggle eyes; or small eyes might have been joined with feet of any other form. What was it therefore which brought them together in the mole? That which brought together the barrel, the chain, and the cogs, in a watch—design; and design, in both cases, inferred from the relation which the parts bear to one another in the prosecution of a common purpose. …
    In a word; the feet of the mole are made for digging; the neck, nose, eyes, ears, and skin are peculiarly adapted to an underground life; and this is what I call relation'.

of God. These days few biologists would try to pin their religious faith, if they have any, on biological evidence, and the apparent design to which Paley referred would be attributed instead to the evolutionary mechanism which Charles Darwin called natural selection.

Charles Darwin in old age. Line drawing adapted from a photograph from *Life and a Letters of Charles Darwin* (1891). Wikimedia, https://commons.wikimedia.org/wiki/File:Charles-darwin-portrait-sitting-on-chair-sketch.png, Public Domain.

Darwin's theory of evolution by natural selection is universally accepted among serious biologists (except for a few so-called creationist scientists), even if arguments continue over the details. Darwin proposed a three-stage cycle that starts with variation in the form and behaviour of individuals. In any given set of environmental conditions some individuals are better able ~~than others~~ to survive and reproduce than others because of their distinctive characteristics. The historical process of becoming adapted notches forward a step if the factors that gave rise to those distinctive characteristics are inherited in the course of reproduction. Suppose, for example, that an individual bacterium happens to have heritable characteristics that make it resistant to an antibiotic. While all the others are killed by the antibiotic, this one will survive and multiply rapidly. Before long, the world is full of antibiotic-resistant bacteria. Darwinian evolution requires no unconscious motives for propagation — let alone conscious ones.

Biologists should not write evolutionary accounts in which the past is seen as leading purposefully towards the goal of the present blissful state of perfection. A clear distinction is necessarily and wisely drawn between the present-day utility (or function) of a biological process, structure or behaviour pattern, and its historical, evolutionary origins. Darwin noted, for example, that while the bony plates of the mammalian skull allow the young mammal an easier passage through the mother's birth canal, these same plates are also present in the mammals' egg-laying reptilian ancestors. Their original biological function clearly must have been different from their current function.

The distinction between current function and historical evolution is all the more necessary because current adaptations may result from the experience of the individual during its lifetime. Human hands form calluses to protect against mechanical wear, and muscles develop in response to the specific loads placed upon them during exercise. Behaviour, in particular, becomes adapted to local conditions during the course of an individual's development, whether through learning by trial and error or through copying others. These are all examples of adaptations that are acquired during the lifetime of the individual, and they are clearly distinct from adaptations that are inherited.

An important advance in thinking was made by the Nobel Prize winner, Niko Tinbergen.[3] He pointed out that a number of fundamentally different types of question may be asked when studying behaviour. 'How does it work?' 'How did it develop?' 'What is it for?' and 'How did it evolve?' In the case of fully-formed behaviour, questions to do with control and function are current, whereas questions to do with evolution and development are historical.

Tinbergen's distinctions can be illustrated with a commonplace example. Suppose drivers are asked why they stop their cars at red traffic lights. One answer would be that a specific visual stimulus—the red light—is perceived, processed in the central nervous system and reliably elicits a specific response (easing off on the accelerator, applying the brake and so on). This would be an explanation in terms of the way in which the traffic light controls the behaviour of drivers. A different but equally valid answer is that individual drivers have learnt this rule by past observation and instruction. This is an explanation in terms of development. A functional explanation is that drivers who do not stop at red traffic lights are liable to have an accident or, at least, be stopped by the police. Finally, an 'evolutionary' explanation would deal

---

3   Tinbergen, N. (1963), On aims and methods of ethology. *Zeitschrift für Tierpsychollogie*, 20, 410–433. An appreciation and an update of Tinbergen's thinking half a century later is given in Bateson, P & Laland, K. (2013), Tinbergen's four questions: an appreciation and an update. *Trends in Ecology & Evolution*, 28, 712–718.

with the historical processes whereby a red light came to be used as a universal signal for stopping traffic at road junctions. All four answers are equally correct, but they reflect four distinct levels of enquiry about the same phenomenon. To use the traffic lights example once again, their efficient regulation of the behaviour of drivers suggest that they have been designed by human agency undoubtedly correct in this case.

The perception that behaviour is designed springs from the relations between the behaviour, the circumstances in which it is expressed and the resulting consequences. The closeness of the perceived match between the tool and the job for which it is required is relative. In human design, the best that one person can do will be exceeded by somebody with superior technology. If you were on a picnic with a bottle of wine stoppered with a cork but had no corkscrew, one of your companions might use a strong stick to push the cork into the bottle. If you had never seen this done before, you might be impressed by the choice of a rigid tool small enough to get inside the neck of the bottle. The tool would be an adaptation of a kind. Tools that are better adapted to the job of removing corks from wine bottles are available, of course, and an astonishing array of devices have been invented. One ingenious solution involved a pump and a hollow needle with a hole near the pointed end; the needle was pushed through the cork and air was pumped into the bottle, forcing the cork out. Sometimes, however, the bottle exploded and this tool quickly became extinct. As with human tools, what is perceived as good biological design may be superseded by an even better design, or the same solution may be achieved in different ways.

The proposition that living organisms' bodies, brains and behaviour were adapted over the course of evolution and by their suitability to the conditions in which they live is familiar to most non-biologists. An adaptation is a characteristic of an organism that makes the organism better suited to survive and reproduce in a particular environment—better suited, that is, than if it lacked the crucial feature.

Design of machines

Within an individual, as well as between individuals, different systems of behaviour are variable both in their development and in their organisation. Some insight into why this should be may be obtained by looking at machines. Tailoring a system to a specified job while building in flexibility is a problem that human designers of machines must face again and again. Robots with even simple forms of regulation do things that look remarkably life-like. Similarly, in a game like chess simple rules can generate games of great complexity.

The difficult challenge for the designers of chess-playing computers is to beat the creativity, flair and imagination of a chess Grandmaster. IBM rose to the chess challenge and started its Deep Blue project in 1989.[4] The Deep Blue computer relied on massive parallel arrays with dedicated hardware and software. It had 256 chess-specific processor chips operating in tandem, each capable of analysing up to three million chess moves every second. The whole array could process 50–100 billion moves in the three minutes allotted for each move. It was also equipped with an enormous database of Grandmaster games played in the previous century.

In the initial stages of the project no attempt was made to mimic human thought. Without any 'psychology' to mess things up, the machine would never get tired or make a silly mistake. It would instead depend for its success on raw computing power and its enormous memory. In one second Deep Blue could search ahead through several hundred million possible moves, while its human opponent, the Russian Grandmaster and one-time World Champion, Gary Kasparov, could analyse only one or two. Kasparov himself admitted that quantity sometimes becomes quality. But he had the compensatory benefits of intuition, judgment and experience.

The World Chess Champion Garry Kasparov played successfully against IBM's Deep Blue computer but was beaten by the next version, Deeper Blue. Photo by Jürg Vollmer (2009), Flickr, https://www.flickr.com/photos/maiakinfo/3858951927, CC BY-SA 2.0.

---

4   Bateson, P. & Martin, P. (1999), *Design for a Life*. London: Jonathan Cape, pp. 96–97.

Compared to computers, humans calculate slowly, but are good at recognising patterns. Chess Grandmasters are much better than novice players at remembering patterns of pieces from real games, but no better at remembering arbitrary patterns. Experience helps them to remember patterns that have meaning and link these with the sequences of moves that have the best pay-off in the long run. The surprising consequence is that humans see traps that lie beyond the search horizon of even an exceedingly fast computer.

In 1996 Kasparov played Deep Blue in a six-game match. Kasparov lost the first game, but then put his human skill to good effect and went on to win the match. He was able to do this because he could adapt his strategy in response to what he discovered were weaknesses in his machine opponent. Deep Blue, on the other hand, could not respond to the overview of its human opponent. IBM rose to the new challenge. Deeper Blue, their 1997 successor to Deep Blue, was faster and smarter.

In particular, it could modify its basic strategy between tournaments in response to the playing style of its human adversary. This time the machine managed, albeit with some difficulty, to win the next match against Kasparov.

These chess matches emphasised how important adaptability is in such competitions. An interesting development has been the cooperation between machine and humans.[5] Amateur chess players coached their laptop computers to explore deeply specific positions using human pattern recognition together with their computers' computational power. The resulting combination overcame in competition the superior chess knowledge of grandmasters and the superior computational power of big computers. Average players with average machines beat the best players and the best machines.

In more practical uses, such as the control of traffic flow by co-ordinating the switching of traffic lights or regulating speed limits, a capacity to adapt to new situations is desirable. Faced with novelty, such systems must not change everything at once. If they did, they would quickly collapse into chaotic malfunctioning. So, as with animals, buffering some aspects of the computer's capacities from change is

---

5   See Shyam Sankar, The rise of human-computer cooperation, TED talk (June 2012), http://www.ted.com/speakers/shyam_sankar

crucial. These essential capacities must continue to function in the same way despite radical changes in input. Life brings many requirements.

Impressive though IBM's Deeper Blue computer was, it was dedicated to one complicated but narrowly defined task—playing chess. Gary Kasparov may have met his match on the chess board, but he was able to do a great many other complicated things of which Deeper Blue was incapable. He could make decisions about chess matches and holidays that he would take years into the future. He could run a complicated social life and allocate time to his main biological appetites, none of which were shared with Deeper Blue. He could feel moved by patriotism or spiritual feelings. He could write books and enjoy music. From time to time, he doubtless reflected on his life and his character.

Conflicts in motivation

Kasparov like every other human and every other animal, had many strands to his life. The systems that are involved in running each of these aspects sometimes seem to be semi-autonomous, usually functioning smoothly together but occasionally coming into conflict. Humans feel the conflict most strongly, perhaps, in times of war, when their craving for leadership and their identification with their own group, tribe or nation conflicts with their peace-time commitments and pleasures and, indeed, perception of their own self-preservation. But everybody feels the pull, on most days of even the most routine life, between incompatible activities. You can't eat and sleep at the same time; you can't have a warm shower and take a walk simultaneously—except perhaps during a cloud burst in the tropics.

Much of animal and human behaviour and physiology operates on the basis that considerable autonomy has seemingly been designed into each behavioural system or organ. Interaction necessarily occurs between them to prevent total breakdown when the different parts pull in different directions. A problem of great interest to engineers has been how far machines should emulate biology, using specialised modules like those in the brain that are dedicated to particular jobs such as recognising faces. How far should the modules be built into separately organised systems, each competing for time when they cannot operate simultaneously? Should a 'boss' allocate priority where it is impossible for two activities to occur at the same time? Or should a decision to express a particular form of behaviour depend on weighting the needs of competing systems?

Designers of intelligent machines often opt for distributed control, known as heterarchy[6] (as opposed to hierarchy), because of the efficiency it brings. The solutions to the problems of running a smart machine, or an individual life, are also found in the management of human organisations. In contrast with traditional hierarchical bureaucracies, modern public and commercial institutions increasingly tend to have structures and organisational cultures that focus on tasks or projects. The emphasis is on getting the job done efficiently, and this is achieved by bringing together groups of people with the right knowledge and skills. Expertise and teamwork are what counts, rather than formal status. The organizational structure tends to be a matrix of project teams rather than a traditional top-down hierarchy. Such management relies on great flexibility and considerable autonomy for each part of the organisation, with exchange of information and competition occurring when the well-being of the whole demands it. The central control over day-to-day work is minimal and the ways in which each team is set up depends on the need.

How do animals achieve comparable solutions in the development and integration of their behaviour? The ultimate arbiter of priority in organising their own behaviour is reproductive success. The consequences of giving priority to this aspect of their biology are sometimes astonishing—at least when judged from a human perspective. The male emperor penguin brooding his mate's egg over the Antarctic winter cannot be relieved by his mate because the growth of the ice shelf puts the sea and food beyond reach. So, in the interests of producing an offspring, he fasts for months—a feat any human would find impossible. Other potential solutions to this problem, such as shorter stints of brooding and trekking repeatedly across the ice shelf during the winter, presumably proved to be less successful. The penguins that fasted all winter were the ones whose ancestors had best survived with this adaptation. Examples like this emphasise how dependent is the organisation of behaviour on the ecology of the species. Differences between individuals in the processes of development are to be expected.

---

6 http://en.wikipedia.org/wiki/Heterarchy

Conclusions

The developmental progression from a single cell to an integrated body of billions of cells, combining to produce coherent behaviour, is astonishingly orderly. Just as animals grow kidneys with a specialised biological function, adapted to the conditions in which they live, so they perform elaborate and adaptive behaviour patterns without any previous opportunities for learning or practice. Particular behaviour patterns are like body organs in serving particular biological functions; their structure is often likely to have been adapted to its present use by Darwinian evolution and by their adaptability. It depends on the ecology of the animal. Their structure and behaviour develop in a highly coordinated and systematic way.

From the standpoint of design, systems of behaviour that serve different biological functions, such as cleaning the body or finding food, should not be expected to develop in the same way. In particular, the role of experience is likely to vary considerably from one type of behaviour to another. In predatory species, such as cats, cleaning the body is not generally something that needs special skills tailored to local conditions, whereas capturing fast-moving prey requires considerable learning and practice to be successful. The osprey snatching trout from the water does not develop that ability overnight. Animals that rely upon highly sophisticated predatory skills, such as birds of prey, suffer high mortality when young and those that survive are often unable to breed for several years because they have to hone their skills before they can capture enough prey to feed offspring. In such cases, a combination of different developmental processes is required in order to generate the highly tuned skills seen in the adult.

Retaining the concept of design brings with it insights that biologists might well not have had without it. Even so, its use generates an unforeseen problem in the current world. The pre-Darwinian ideas about intelligent design have been taken up by the creationists in their attempts to disguise their beliefs as a form of science. As a result all sorts of unpalatable associations are brought up in the minds of biologists when they hear the word 'design'. In attempts to make accessible complicated processes intelligible, various devices are used—like attributing metaphorical intentions to genes (see Chapter 8) or to the weather. These linguistic devices are easily misunderstood. The take home message is, then, that when using a term like design, which means different things to different people, great efforts must be made to ensure the language is not taken too literally. The design in biology is only apparent, but shorn of its religious connotations, understanding the relations between the parts of an organism remains as useful as ever.

2. Imprinting and Attachment[1]

Imprinting provides a striking example of the way in which a particular experience has a specific effect only when the animal is at a certain stage of behavioural development. Indeed, the regulation of imprinting predisposes many species of bird to learn the characteristics of their parent at what would appear to be the biologically appropriate time in their life cycles. It is a good example of how behaviour gives the appearance of being well designed to serve the needs of the young birds.

A Mallard Duck hen calls vigorously as she leads her ducklings who have already formed an attachment to her. Photo by Crystal Marie Lopez (2010), Flickr, CC BY-ND 2.0, https://www.flickr.com/photos/labellavida/4697991484

Even though birds like domestic chicks and mallard ducklings, the species most commonly used in studies of imprinting, respond to a wide range of objects before they have formed an attachment, they

---

1 Much of this chapter is based on an updated version of Bateson, P. (1973), Internal influences on early learning in birds. In: R.A. Hinde and J. Stevenson Hinde (eds.), *Constraints on Learning: Limitations and Predispositions*. London: Academic Press, pp. 101–116, with thanks to the Master and Fellows of St John's College, Cambridge.

respond much more strongly to some than to others. This selective responsiveness is a major constraint on what is readily learnt in the imprinting situation. The characteristics of the stimuli that are most effective in eliciting social behaviour in naïve birds vary from species to species. In general stimuli that resemble most appropriate biological objects are preferred by naïve chicks and ducklings more strongly than those that don't.[2]

One feature of imprinting is its apparent restriction to a brief period early in life. At one time it was supposed that a window opened on the external world and then closed again. While the window was open the young animal was affected by certain types of experience and at other times it was not. This interpretation did not follow from the evidence. While maturational changes, occurring independently of specific experience, have been implicated in its onset,[3] the sensitive period is brought to an end by a specific type of experience. Birds become familiar with their immediate environment, whether this be their mother, other chicks, or even the walls of their isolation cage, and come to discriminate between such stimuli and other things that are novel to them. When they can tell the difference, they avoid the strange object and subsequently

---

2 Day-old domestic chicks trained with a flashing, rotating light or with a rotating stuffed jungle fowl, the ancestral species of domestic fowl, and then given a choice between them did not differ in their preferences. The stuffed jungle fowl became more attractive than the box by the second day after hatching. The shift towards a stronger fowl bias was also apparent in birds that had been imprinted with either a fowl or a box. Features of the jungle fowl that make it especially attractive as the predisposition emerges are located around the head. They are not specific to jungle fowl since the heads of a stuffed duck and small predator were equally attractive. Under laboratory conditions, the necessary feature detectors for head and neck evidently take longer to develop than do the ones driven by flashing lights and movement. Johnson, M.H. & Horn, G. (1988), Development of filial preferences in dark-reared chicks. *Anim. Behav.* 36.3, 675–683, http://dx.doi.org/10.1016/S0003-3472(88)80150-7. Team of scientists (Vallortigara, G., Regolin, L. & Marconnato, F. (2005), Visually inexperienced chicks exhibit spontaneous preference for biological motion patterns. *PloS Biol.* http://dx.doi.org/10.1371/journal.pbio.0030208), found that animation sequences of point-light-displays in which a few light points are placed on the joints of a digitalized image of a moving hen were more attractive to naïve chicks than the same points of light upside down. The spatial relational properties of the imprinting object have proved to be important (Martinho. A. III & Kacelnik, A. (2016), Ducklings imprint on the relational concept of 'same or different'. *Science* 353.6296, 286–288, https://doi.org/10.1126/science.aaf4247).
3 Experience before hatching is important. When the unhatched chick starts to vocalise, its calls facilitate the preference for the maternal call after hatching (Gottlieb, G. (1988), Development of species identification in ducklings: XV. Individual auditory recognition. *Devel. Psychbiol.* 21.6, 509–522, https://doi.org/10.1002/dev.420210602).

show no evidence of having developed a preference for it. The end of the sensitive period does not mark the point at which learning is complete; it merely marks the point at which the young bird is able to discriminate between stimuli that it has already experienced and other objects.[4]

Imprinting is an example of tightly constrained learning. Paradoxically, its general interest lies in its particularity. The predispositions to respond to particular features and give particular responses to the stimulus are central in the case of imprinting. Processes that change as a result of experience are dependent on features that have developed before imprinting has taken place. In other examples of learning that have different functions and are involved in different motivational systems the inter-dependence is less obvious, but present nonetheless. The differences in the ways in which animals learn can be explained in terms of variation in the perceptual and motivational mechanisms used in the various contexts in which learning occurs. In general, the properties of the whole animal allow for the evolution of differences in function.

Attachment in humans

Analogies between imprinting in birds and the development of attachments in humans have been drawn, particularly by the great psychiatrist John Bowlby.[5] The day-old baby is affected by her auditory experience before birth and she prefers the sound of her mother's voice to that of other women. She has a clear predisposition to respond to face-like images and rapidly develops a preference for the details of her mother's face. She makes much effort to maintain contact with her mother and is upset when the behavioural exchange with the mother is disrupted.[6]

---

4   Bateson, P. (1979), How do sensitive periods arise and what are they for? *Anim Behav.* 27.2, 470–486, https://doi.org/10.1016/0003-3472(79)90184-2. For a more recent review of sensitive periods in the development of brain and behaviour see Knudsen, E.I. (2004), Sensitive Periods in the Development of the Brain and Behavior. *J. Cogn. Neurosci.* 16.8, 1412–1425, https://doi.org/10.1162/0898929042304796
5   Bowlby, J. (1969), *Attachment and Loss. Vol. I: Attachment.* London: Hogarth Press. Bowlby was concerned to provide an empirical basis to the field of psychoanalysis.
6   The elegant work of Murray, L., & Trevarthen, C. (1986), The infant's role in mother-infant communications *J. Child Language* 13.1, 15–29, https://doi.org/10.1017/s0305000900000271 showed that a two month old baby would respond normally to the face of her mother on a TV screen but was upset when a time delay was inserted between her behaviour and that of her mother.

The dynamics of her social relationships as she develops is the subject of much research. In this respect the work on imprinting in birds and the development of social attachments in children have diverged. The work on imprinting in birds has been focused on those species that are feathered and active in early life, with particular attention paid to the detailed mechanisms involved. The work on attachment processes in humans has focused on the ramifying consequences of the child's experiences on her subsequent behaviour.⁷ As so often happens, the bodies of knowledge have separated and attempts to bring them together have often been at a superficial level. Nevertheless, the general conceptual questions have value inside the various silos of knowledge.

The human mother and her child have formed a strong attachment to each other. Photo by Bob Whitehead (2006), Flickr, https://www.flickr.com/photos/kryten/125710155, CC BY 2.0.

Imprinting in the wild

The conditions under which imprinting is studied in the laboratory are necessarily impoverished and artificial. The results can give a seriously misleading view of what happens in the wild. Chicks and ducklings spend most of the daylight hours on the first day after hatching being brooded by their mothers. The little birds hardly seemed to pay her any attention. Their activity around the hen does increase substantially on the second day after hatching, or even later if the ambient temperature is low.

Although the development of new preferences is initially prevented by escape from novelty or by the low level of social responsiveness to

---

7   Holmes, J. (2010), *Exploring in Security: Towards an Attachment-informed Psychoanalytic Psychotherapy*. London: Routledge, https://doi.org/10.4324/9780203856321

unfamiliar things, enforced contact may wear down these behavioural constraints to the point where the bird does develop a new preference. This flexibility could be of some functional importance in colonial nesting species such as gulls. In the absence of parents, for which the young bird forms its strongest preference, the bird may still be able to survive by responding socially to other adults and inducing them to feed it.

Even in the laboratory, when a recently hatched mallard duckling or domestic chick, which has been sitting quietly in a dark incubator, is removed and alley at room temperature, it soon begins to move about. Before long it starts emitting shrill peeps, often referred to as 'distress' calling, and it shuffles about in disorientated fashion with its neck extended. If a conspicuous visual stimulus is now presented to the bird, it orientates towards the stimulus and its distress calling stops. In many ways, its behaviour resembles that of a bird that has become separated from its mother, vigorously searching for her.

Such an observation suggests that even before they have been imprinted, the bird will behave in a way that increases the likelihood of their making visual contact with their parent or a surrogate.[8] The animal plays an active part in determining the kinds of things that it will learn and will continue to do so even after the imprinting process is under way. The bird cannot predict what the back view of its mother is like from knowledge of her front view. If a bird that has formed an attachment to an individual can respond selectively to that individual regardless of its orientation, then the bird must have been exposed to all those views of the parent that it can subsequently identify. It has built up a composite picture of its parent's characteristics. In the normal course of events, the mother will probably present many different aspects of herself during the attachment process while the young are learning her characteristics. Assurance would be made doubly certain if, after learning a certain

---

8   If stimuli that are highly effective in the imprinting situation do bring such searching behaviour to an end, they might be expected to reward the young bird. Naïve domestic chicks and wild mallard ducklings taken from a dark incubator quickly learn to operate a pedal that turns on a flashing rotating light. Age, and prior experience, affect the ability of domestic chicks to learn the pedal-pressing task in the same way as they affect the imprinting process (Bateson, P. & Reese, E.P. (1969), The reinforcing properties of conspicuous stimuli in the imprinting situation. *Anim. Behav.* 17, 692–699).

amount about her, the young actively worked to present themselves with a different view.[9] The active element in the young bird's behaviour makes the attachment process much more flexible and adaptive than it would have been if the bird had simply locked on to the first thing it saw and attempted to maintain contact with that and nothing else.

The incisive, single-shot image conjured up by the term 'imprinting' does not adequately represent what happens. Clearly, acquisition of the complex pattern recognition involved in detecting a particular parent or surrogate from many different angles and distances takes some time. Imprinting with two objects presented in rapid alternation can have a retarding effect on rewarded discrimination learning.[10] It is as though the stimuli are classified together and come to share the same identity. This could be an integral part of the imprinting situation where the young animal has to build up a composite picture of its parent as it obtains the opportunity to view the parent at various angles.

The advantages of doing this are not restricted to the attachment process. Classification together of physically different stimuli may well be necessary for some of the more complex examples of 'concept formation', even though abstraction of common features of different stimuli and generalisation from familiar to novel stimuli are also likely to be involved. The process may also play a larger part in human perception than personal experience suggests — introspection being a poor guide to the distinctions ignored in existing classifications.

Individual recognition

Filial imprinting and sexual imprinting have certain things in common even though sexual imprinting takes place later in development than filial imprinting.[11] Both filial and sexual imprinting have evolved to enable birds to recognise their close kin, but the necessity for kin recognition is different in young and adult. The young bird needs to

---

9   Jackson, P.S. & Bateson, P. Imprinting and exploration of slight novelty in chicks. *Nature*, 251.5476, 609–610, https://doi.org/10.1038/251609a0
10 Bateson, P. In Heyes, C. & Huber, L. (eds.), *The Evolution of Cognition*. Cambridge, MA: MIT Press, 2000. pp. 85–102.
11 Vidal, J.-M. (1980), The relations between filial and sexual imprinting in the domestic fowl: Effects of age and social experience. *Anim. Behav.*, 28.3, 880–891, https://doi.org/10.1016/s0003-3472(80)80148-5

discriminate between the parent that cares for it and other members of its species because parents discriminate between their own offspring and other young of the same species, and may attack young that are not their own. Adult behaviour of this kind is well known in many mammals and birds. In most cases the parent that cares exclusively for its own young will be more likely to rear them to independence than a parent that accepts and cares for any young that come up to it. The suggestion is, then, that filial imprinting is required for individual recognition of parents and is a secondary consequence of the evolutionary pressures on parents to discriminate between their own and other young. In each generation individuals may differ in the stage of development when their filial responsiveness to parent-like objects first increases. Those that do it too early obtain inappropriate or insufficient information about their parents. They might, for instance, have inadequate opportunities to explore all facets of their parent and so fail to recognise it quickly enough later on when quick recognition is important. Those that do it too late respond in a friendly way to hostile members of their own species and consequently suffer attacks. In these different ways the optimal timing for the increase in intrinsic responsiveness could have evolved. It would be critically affected by how rapidly the parents learn to discriminate between their own young and other young.

The evolutionary pressures that give rise to sexual imprinting are likely to have been quite different. Sexual imprinting enables an animal to learn the characteristics of its close kin and subsequently choose a mate that appears slightly different (but not too different) from its parents and siblings (see Chapter 7).

Conclusions

Imprinting is an example of tightly constrained learning. The predispositions to respond to particular features and give particular responses to the stimulus are central to understanding what happens. The robust processes of development make possible the plastic changes in behaviour that follow. Processes that change as a result of experience are dependent on features that have developed before imprinting has taken place. In other examples of learning that have different functions and are involved in different motivational systems the interdependence is less obvious, but present nonetheless. The differences in the ways in which animals learn can be explained in terms of variation in the perceptual and motivational mechanisms used in the various contexts in which learning occurs. In general, the properties of the whole animal allow for the evolution of differences in function. Imprinting is a good example of how bringing together all the factors known to affect it provides a systems approach to development. It also has the appearance of being well designed for the needs of the animal.

3. Rules and Reciprocity[1]

The study of development has attracted some of the most bitter and protracted controversies in the whole of psychology and ethology. The arguments reflected more general ideological battles about nature and nurture. Consequently, much research was concerned merely to establish that a particular kind of internal or external factor could be important, or that evidence could be obtained for a certain logical possibility. In recent years, such activity has abated with the growing acceptance that both internal and external factors can play important roles in the development of any one pattern of behaviour. Also, the air has been cleared by the realization that an interest in how behaviour has been adapted to its present uses is not the same as an interest in what makes one individual animal different from another one.

Two radically different models have been proposed for what is happening when behaviour develops. On one view straightforward correspondence can be found between genes and innate behaviour on the one hand and between the environment and learned behaviour on the other. The word 'innate' has many different meanings attached to it: present at birth; a behavioural difference caused by a genetic difference; adapted over the course of evolution; unchanging throughout development; shared by all members of a species; and not learned. 'Instinct' is deployed in similar ways to innate. When the justification for using one of the meanings of innate or instinctive has been demonstrated it does not follow that another of the meanings will

---

[1] Parts of this chapter were taken, with permission, from Bateson, P. (1976), Rules and reciprocity in behavioural development. In: P. Bateson & R.A. Hinde (eds.), *Growing Points in Ethology*. Cambridge: Cambridge University Press, pp. 401–421.

necessarily apply. Even if a behaviour pattern develops without obvious practice or example, it may subsequently be modified by learning. For instance, a blind baby may start to smile in the same way as a normal baby. But that does not mean that later on in their lives, sighted people will not modify their smiles to expressions that are characteristic of their own culture. A classification of behaviour into innate and not innate merely causes confusion.[2]

The second view of development does not recognize the distinction between behaviour that is not learned and behaviour that is acquired. It proposes instead that as the animal develops, it is not merely affected by its genes and its environment. The animal's state influences which genes are activated from time to time and the animal also alters the character of the environment as it develops.[3] While such transactions of this kind seem reasonable, this second model is often perceived as being too complicated and too vague. Such objections start to fall away as the nature of developmental processes are unravelled. Gradually scientists have become aware that what is needed is an approach that will cope with the multiple and variegated nature of the factors that make individuals different from each other and the interactions that take place between those factors. This amounts to a systems approach.

Models of development

In a helpful visual aid to the biologist who has difficulty in grasping the abstractions of a mathematical model, Waddington[4] represented the development of a particular part of a fertilised egg as a ball rolling down a tilted plane which is increasingly furrowed by valleys. He called the surface down which the ball rolls the 'epigenetic landscape'. The essential point is that the mounting constraints on the way tissue can develop are represented by the increasing restriction on the sideways movement of the ball as it rolls towards the front lower edge of the

---

[2] A discussion of the concept of innateness is given in Mameli, M. & Bateson, P. (2006), Innateness and the Sciences. *Biol. Philos.*, 21.2, 155–188, http://dx.doi.org/10.1007/s10539-005-5144-0

[3] Lehrman, D.S. (1970), Semantic and conceptual issues in the nature–nurture problem. In: Aronson, L. Tobach, E. & Rosenblatt, J.S. *Development and Evolution of Behavior*. San Francisco: Freeman, pp. 17–52.

[4] Waddington, C.H. (1957), *The Strategy of the Genes*. Allen & Unwin: London.

landscape. The landscape represents, therefore, the mechanisms that regulate development.

Waddington's model is attractive to the visually minded because it provides a way of thinking about developmental pathways and about the astonishing capacity of the developing system to right itself after a perturbation and return to its former track. To take a specific example from post-embryonic development, if a juvenile rat is starved during its development, its weight falls while it is being deprived. When it is put back onto a normal diet, its weight curve rapidly picks up and rejoins the growth curve of the rat that has not been deprived. Similar examples of growth spurts after illness are well known in humans. For the moment the possibility that the individuals showing the catching-up phenomenon may differ in undetected ways from normal individuals can be ignored. The prime question is how weight gain is controlled and how two individuals with different dietary histories end up weighing the same.

The systems theorists have laid considerable emphasis on the self-correcting features of development, and have called the convergence of different routes on the same steady state 'equifinality'.[5] Waddington's epigenetic landscape suggests a way of handling equifinality. If the ball rolling down the landscape encounters an obstacle in one of the valleys and is not stopped dead, it will ride up round the obstacle and fall back into the valley down which it had been rolling.

Waddington's model is, of course, informal and he would have been the first to point to its limitations. It is not difficult to simulate with greater rigour a system that compensates for short periods of food deprivation during development. If the amount of food an animal attempts to eat is determined by a comparison between a predetermined setting and the actual weight of the animal and if the value of the preferred weight is increased as the animal ages, a model similar to biological reality can be obtained. To make things more realistic the predetermined increments in the preferred value first increase and then decrease as the hypothetical animal gets older.

By arranging for the preferred value of the closed feedback loop to be changed according to some predetermined plan, the system has

---

5   Capra, F. & Luisi, P.L. (2014), *The Systems View of Life*. Cambridge: Cambridge University Press, http://dx.doi.org/10.1017/cbo9780511895555

been made interactive. In the simplest case the developmental process is essentially ballistic — its pathway is determined in advance and does not depend on a dynamic interaction between the system and other factors that might change during the course of development. Even rather simple explanations can account for different developmental routes leading to precisely the same steady state. The phenomena, which were so entrancing to an old-fashioned vitalist, do not pose inordinate conceptual problems.

Children differ astonishingly in the age at which language development begins. Some may begin before the end of their first year and others may not utter a recognisable word until they are three or more. Furthermore, during acquisition, styles of forming word patterns may be markedly different. Despite these enormous differences, it is remarkably difficult to pick out the early developers when the children are older. Put cautiously, behaviour at one stage of development is an exceedingly poor predictor of behaviour at another. Put more boldly, a child which has been initially slow to develop can demonstrate the catch-up effect seen in tissue growth and reach the same steady state in one aspect of language ability as a more precocious child.

This example from language acquisition can be matched by many others from child development, but it is sufficient to demonstrate not only the advantages but also the difficulties of employing the concept of equifinality in developmental studies of behaviour. Despite practical and philosophical difficulties, achieving equifinality in behavioural development does not pose insuperable problems of principle. Explanations for the control of weight can be readily adapted to behavioural examples. The preferred value against which the actual state is compared can be for, say, the proprioceptive feedback from a certain action or, at another level, the feedback provided by the behaviour of a parent. The justification for thinking in these terms is that it provides a different perspective from the more conventional interactional approach and suggests new ways of looking at the evidence.

Even a simple way of generating convergence could lead to marked differences in the pattern of development even though the final outcome was the same. Just as two rats with different food preferences can put on weight at the same rate, so different types of action can lead to the same behavioural end-point. A feature of a system dependent for its control on feedback is that it need not be fussy about how a match between the actual value and the preferred value is achieved. It is the consequences of an action that counts, not the precise form and patterning of that action. Admittedly, possible courses of action may be so constrained that the system is likely to do only one thing when a mismatch between the actual state and the preferred value is detected. The constraints need not be great. For example quite different combinations of muscles in a locust's leg contract to produce the same overall movement of the leg — the explanation being that movement is controlled by means of sensory feedback.[6]

Alternative pathways

Different routes to the same goal may be achieved even more dramatically than in the cases already considered if the young individual is equipped with two or more alternative systems controlling development of the same pattern of behaviour. Redundancy of this kind is common enough in man-made machines when lives are at stake, as in an airliner. Clearly, redundant developmental systems could be highly adaptive for an individual, particularly if the alternative control systems were matched to different environmental conditions to which they were appropriate — the provision of special horses for particular courses. The existence of other systems protects against failure, but from time to time individuals are faced with the situation where no amount of tactical manoeuvring will enable one of their developing systems to proceed along a particular route. Such an individual is a bit like a traveller who arrives at a station only to find that the trains have been cancelled. He or she can still reach their destination but only by choosing a different method of getting there.

If contingency arrangements of this kind have been adapted during evolution, Waddington's epigenetic landscape would have to be redrawn so that some valleys ran together again. It could be argued that a ball that had descended by one valley had had a different history from one that had descended by another so that even though the balls ended

---

6 Hoyle, G. (1970), Cellular mechanisms underlying behavior-neuroethology. *Adv. Insect Physiol.* 7. 349–444, http://dx.doi.org/10.1016/S0065-2806(08)60244-1

up in the same place, the concept of equifinality was valueless. This answer would depend on whether the different histories did indeed leave distinctive traces on the metaphorical ball. Even if they did, the objection might still not be serious since the resulting differences might be biologically trivial by comparison with the ultimate similarities.

In other contexts, inputs that may be relatively non-specific are frequently required to facilitate the development of particular systems. The inputs may be provided by external environmental conditions or by feedback from the animal's activities such as its own vocalisation. It may not yet be possible to give a clear instance where different developmental control mechanisms generate the same behavioural end-product.

The biological function of some of the behavioural mechanisms found in many developing animals, particularly higher vertebrates, is the gathering of information. Their predispositions to learn the characteristics of certain things can be highly specific. Such proclivity can be extremely important in directing the course of development. A good example is provided by the active role of the young precocial bird in imprinting (discussed in Chapter 2). This example also illustrates the more general point about modifiability of control mechanisms.

Another kind of modification dependent on environmental conditions is suggested by the stunting of growth if animals or humans are starved for long enough during development. The simple model used for the control of weight can be readily adapted to cope with such evidence by making the extent of the increments in preferred weight dependent on the size of the increments between the preferred weight and the actual weight. If the discrepancy is large the increment in preferred weight is modified so that it is less than if the discrepancy is small. This simple rule, which could be specified in advance, would greatly enhance the dynamic interaction with the environment. It would have one interesting consequence that would be particularly striking if the normal growth curve were sigmoid with the period of maximum growth occurring mid-way through development. The stunting effects of starvation would be particularly marked at times of rapid growth. This would give rise to periods in development when the animal was especially vulnerable to environmental disturbance.

Rules for changing the rules

The biological advantage of a rule that allows for a change in the preferred value is that the animal does not endlessly attempt to reach a state that may never be achievable in the particular conditions in which it is developing. That the conditions for the development of one system are less than optimal does not imply that conditions are bad all round; normal development of the individual's other systems may still be possible. Although it may be handicapped, its chances of surviving and leaving offspring may not be reduced to nothing.

One behavioural example of settling for less than the best is the nest-site choice of the blue tit. In the spring the tits visit a large variety of crannies many of which are obviously unsuitable. One way of interpreting their behaviour would be that, if the actual site did not match up to the characteristics of an optimal nest-site, they kept searching — to begin with at least. If optimal sites were unavailable or already occupied, the birds would ultimately nest in places they had previously rejected. It would make good sense if they were equipped with a rule for gradually relaxing the conditions under which searching for a nest-site was brought to an end and nest building began. Once the bird has selected a sub-optimal site it will, for that breeding season at least, prefer it even if an optimal site should subsequently become available.

The modification of preferred values can be examined in the context of emerging social relationships. Suppose that it is important for the maintenance of a relationship between two individuals that they both have the same general pattern of behaviour — the same activity rhythm, for example. In the early stages of a relationship differences in pattern might well exist but these might reflect nothing more than the relatively unimportant peculiarities of personal history.

It might be possible for one or both of the partners to change their preferred patterns without cost. If a pattern of behaviour is achieved by comparison with a preferred standard, that same standard could also be used for judging a companion's behaviour. Individual A's standard could be changed by individual B's and vice versa. Any mismatch would lead to the individuals breaking-off contact with each other. It would therefore be necessary to provide for a mechanism that would, in the early stages of a developing relationship, over-ride or inhibit the consequence of a mismatch. For example, two individuals might be drawn together by the physical appearance of each other. During the 'honeymoon' period the relatively subtle differences of behaviour would be ignored. It would only be later, when the effects of physical appearances had started to wane that a mismatch of behaviour would become important and lead to a disruption of the relationship. In the intervening period one or both of the individuals could have changed its pattern of behaviour so as to correspond to that of the other. The flexibility of an individual might be constrained by some social roles and facilitated by others.

An example of such behavioural meshing comes from observational studies of the relationship that develops between mother and infant rhesus monkeys.[7] Independent measurements were made when the mother left her infant and when the infant left its mother. In some pairs the probability that the mother would leave the infant at any particular moment after they had come together was closely related to the probability that the infant would leave the mother. Such meshing could, of course, be obtained in a variety of ways. For example, the two individuals might become highly sensitive to the immediate cues provided by their partner. If other things are equal, and if apparent plasticity of preference is not merely elasticity, then the pattern of behaviour should be maintained for some time in the absence of the particular partner with which the pattern developed.

Mother rhesus monkey with her offspring. Photo by Laszlo Ilyes (2007), Flickr, CC BY 2.0, https://www.flickr.com/photos/laszlo-photo/495498455

---

7   Hinde, R.A. & Simpson, M.J.A. (1975), Qualities of mother-infant relationships in monkeys. *Ciba Foundation Symp.*, 33, 39–67, http://dx.doi.org/10.1002/9780470720158.

So long as a rule for changing a rule develops reliably in the individual, the outcome of Darwinian evolution is indifferent to how that came about. The outcome may have been arrived when certain environmental conditions were invariant and reaching adulthood successfully may depend on the maintenance of those conditions.

Coordination in development

If internal mechanisms have developed, by some means or another, to control later stages of behavioural ontogeny, a considerable degree of coordination is likely to exist between different mechanisms. For example, the rates of development of two patterns of behaviour may be independently influenced by interactions with the environment; it may be important that the development of one does not outrun the development of the other. Alternatively, the order in which behaviour patterns develop may be important; for example, exploration of the environment may be disastrous if it occurs before a young animal has established some standards of what is familiar. In such cases acquisition of information must precede performance. Once a motor pattern producing the appropriate feedback has been established, dependence on feedback can be reduced or even eliminated and the animal can accelerate the output rate. This is a bit like a musicians learning a new part. While they are able to monitor the individual sounds they are making to ensure their accuracy, they must allow enough time between notes. In the final performance when such control is no longer needed, the gaps between notes can be reduced.

The processes involved in plasticity can operate at many different levels, ranging from the molecular to the behavioural, some involving adaptability to what may be novel challenges and some responding conditionally to local circumstances. The results of development variation can be triggered in a variety of ways, some mediated through the parent's characteristics. Sometimes variation arises because the environment triggers a developmental response that is appropriate to those ecological conditions. Sometimes the organism makes the best of a bad job in suboptimal conditions. Sometimes the buffering processes of development may not cope with what has been thrown at the organism, and a bizarre set of characteristics is generated. Whatever the adaptedness of the characteristics, each of these effects demonstrate how a given genotype will express itself differently in different environmental conditions.

The contrasting properties of resistance to change and changeability — of elasticity and plasticity — are often found within the same material object. Stretch a metal spring a little and it will return to its former shape. Stretch it too far, however, and it will permanently take on a new shape. Adult humans, too, exhibit plasticity as well as elasticity in their values and personalities; they remain recognisably the same individual in a variety of situations, yet retain the capacity to change (see Chapter 4). Compare the robustness of most people in response to life's buffetings with the way that some individuals profoundly modify their behaviour and attitudes. Continuity and change are not incompatible. The brains that generate behaviour do not consist of springs, of course, but the general property of getting back on track coexists with an ability to alter direction.

The implication of examples such as these is that when certain conditions have been satisfied, new mechanisms of control can be brought into operation. In self-modifying systems, for instance, the conditions necessary for progressing to the next stage of development could be the levelling-off of modification — in other words, the achievement of a steady state. This type of explanation would side-step an unprofitable debate about the precise chronology of developmental stages. It would focus attention on the environmental conditions and on the state of the individual associated with a transition from one stage of development to the next rather than on age as such.

Conclusions

The two points of view alluded to at the beginning of this chapter that have sometimes seemed incompatible can be made compatible. Far from being irreconcilable, the approaches of theorists interested in interactions and those interested in control mechanisms usefully complement each other. In brief, the development of behaviour often requires internal rules for its guidance, but reciprocity between the organism and its environment is also needed in order to give those rules greater flexibility and definition.

Plasticity in response to different environmental conditions may often usefully reside in those mechanisms that determine action by matching actual input values with preferred values. The main point is that if individuals have rules by which their behaviour is controlled, functional reciprocity between the developing individual and its environment can be usefully achieved by equipping the animal with rules for changing the rules. This feature of development has all the appearance of being well designed.

4. Discontinuities in Development[1]

Many aspects of body and behaviour change markedly during the course of development, sometimes quite suddenly. Discontinuous change is most obvious during the first two decades of a human life, for example at birth and puberty. Such discontinuities are not mysteries. Many physical and biological systems are capable of changing in an abrupt, discontinuous way. Steadily increasing the pressure on a light switch does not produce a steady increase in the brightness of the bulb it controls. The switch has a point of instability, so that one moment the bulb is dark and the next moment it is fully lit. Similarly, a relatively small internal or external change can quickly transform a developing organism's characteristics to something that looks quite different. The fertilised egg of an animal rapidly divides becoming a ball of cells, the blastula. The cells continue to divide, but do so at slightly different rates. The steady change is such that the blastula suddenly seems to collapse on one side like a deflated rubber ball and a two-layered structure called the gastrula is formed. The embryo has changed its appearance dramatically as a result of a process of continuous growth.

Sudden changes in behaviour during an individual's development may have biological utility, reflecting the changing ecology and needs of the individual as it gets older. The relatively abrupt alteration in the method of feeding at weaning, or in the mode of behaviour towards members of the opposite sex at puberty, are obvious enough. Becoming

---

[1] This chapter is drawn in part, with the publisher's permission, from Bateson, P. (1978), How does behavior develop? In: P. Bateson & P.H. Klopfer (eds.), *Perspectives in Ethology* Vol. 3. *Social Behavior*. New York: Springer, pp. 55–66. Some animal examples are given but the chapter focuses particularly on discontinuities in human development.

an adult involves a long process, and at each stage in development the individual must cope with challenges of the particular world in which it is living at the time. The caterpillar feeds voraciously and has adaptations for doing so, and by its camouflage or behaviour avoids as best it can the attentions of other animals that would eat it. Then its body goes through a dramatic change while a pupa and it emerges as a butterfly. Such metamorphoses have no exact parallel in humans but the requirements of living in a very different world from that of an adult are real. Are the apparent discontinuities seen in development related to the changes in function?

Monarch butterfly before and after metamorphosis. Monarch butterfly caterpillar. Photo by Antilived (2006), Wikimedia, https://commons.wikimedia.org/wiki/File:Monarch_catepillar.jpg, CC-BY 3.0. Monarch butterfly. Photo by LyWashu (2008), Wikimedia, https://commons.wikimedia.org/wiki/File:Danaus_plexippus_01.jpg, CC-BY 3.0.

In the behavioural literature, 'discontinuity' is used in several different senses. Strictly, it should refer to a sudden alteration in the characteristics of a system. So, for instance, when kittens suddenly start to take solid food in appreciable quantities, their rate of weight increase shows a marked rise since, before it happened, the mother was no longer able to provide sufficient milk to sustain the needs of her growing offspring.[2] 'Discontinuity' is also used for the reorganization of rank on a given measure, so that when scores of attentiveness, vocalization, and sniffing in 2-month-old human infants fail to predict their scores on the same

---

[2] Martin, P. & Bateson, P. (2007), *Measuring Behaviour*. 3rd ed. Cambridge: Cambridge University Press.

measure at ~~4~~ four months, a discontinuity is said to have occurred. It is obvious that a discontinuity in the first sense of a change in organisation does not necessarily imply a discontinuity in the second sense of the rank ordering of individuals. The rank order of the kittens in terms of weight could remain unchanged with one kitten remaining the heaviest throughout development. In fact, the sudden change occurs at different ages in different litters, so the rank order may be altered, but only temporarily until some time after all the kittens have started to eat solid food.

Another type of evidence for discontinuities comes from longitudinal correlation studies in which the same pattern of behaviour is measured repeatedly in a group of subjects through development. Behaviour at one stage may successfully predict behaviour at the next and the next but not the one after that. So long as the rank order is dependent on new sources of variation or fluctuates randomly, extreme values tend to regress toward the mean. As a result, correlations between behaviour at one stage and behaviour at succeeding stages tend to diminish with the lapse of time. Even so, a particularly rapid reduction in the strength of a correlation requires additional explanation.

Associations between different measures side-step the problem of whether new behaviour patterns really are new. But it creates a fresh problem, which is whether the 'old' behaviour patterns really are the same when measured at later stages of development. The superficial descriptive similarities between different behaviour patterns may be deceptive. A mistake of this type might be especially likely when the measure in a standard test was something like 'Latency to approach' or 'Time near object'. Careful observation of the descriptive structure of the motor patterns might indicate that quite different systems of behaviour are involved at the different stages of development. For instance, when domestic chicks approach a novel visually conspicuous object, they may be responding socially at one day old but asocially exploring it when they are one week old. The character of the discontinuity can be investigated by examining not only the detailed form of the behaviour but also the consequences for the animal of its response before and after the change.

Changes in rank order need to be handled with particular care because the changes can usually be explained in a variety of ways, some of which are trivial. Positive correlations between a measure taken at one stage of development and those taken at others may evaporate for an uncomplicated reason, namely, that the measurements start to clump together. This clumping might arise because the measurements reach some ceiling (or floor) on a scale of measurement. For instance, a correlation between the performances of children on a simple test of addition at one age and their performances on the same test at later ages might drop to zero because a point is reached when all the children get full marks.

An absence of a correlation between two measures as determined by a straightforward statistical test need not necessarily mean that no relationship exists. A change from a linear to an inverted U-shaped relationship might easily be misinterpreted as a loss of any relationship. Consider a case in which IQ is repeatedly measured at different times in a group of children. Suppose that the test is one that is especially appropriate for a certain age range and contains questions like 'Continue the series O,T,T,F,F...' Bright children have no difficulty in providing the answer: 'S,S,E...' Bright adults rack their brains looking for complex rules that might provide some meaningful sequence and totally miss the point that the series consists of the first letters of One, Two, Three, Four, Five... Children who do well on an IQ test at one stage switch to a more adult-like problem-solving strategy before their peers and consequently perform relatively badly on what is for them an inappropriate test. So an inverted U-shaped relationship between cognitive ability at the two stages might easily arise. The change in the children's strategy would be interesting but would not signify that a fundamental *reordering* of the children had occurred as a result of some event in development. Clearly, this possibility can be examined by plotting the data in graphical form. Furthermore, with appropriate measures of behaviour at later stages in development, it should be possible to show that the original rank order is eventually regained.

In a cohort of individuals growing older together, their rank order on a particular measure at Stage 1 might correspond closely to their rank order at Stage 3. For instance, the factors responsible for the difference in their height at Stage 1 might be the same set of factors responsible for the difference in their height at Stage 3. A quite different set of factors might be responsible for the timing of a growth spurt. Consequently, the rank ordering of height at Stage 2 during the period of the growth spurt might be quite different from that of Stage 1 and Stage 3. Such an effect should be detectable from the data, provided of course that the cohort has been sampled at appropriate points during development.

Changes in development take many forms, which invite different explanations. A qualitatively new pattern of behaviour may appear and an old one may disappear; switching from suckling to eating solid food during weaning is an example. An individual child's consistent tendency to cry more than it laughs might change at a particular age. The rank-order of individuals at one age often fails to predict the order at another age, so an infant who is, say, more attentive than others at two months may be less attentive than others at four months. Sometimes, the slow starter catches up and overtakes the more precocious child. Parents who are proud that their child is reading at the age of four should not assume that the child will turn out to be a genius. Faster development does not necessarily mean a superior outcome.

Loss of continuity

Continuity from one age to the next may be lost for many reasons. One is that development is affected by many influences, not all of which are the same for everybody. Another is that children are often profoundly influenced by the social situation in which they find themselves.

Continuities across age may also be lost temporarily because different children pass through a particular transition at different chronological ages. People who are tall for their age when they are two years old are also highly likely to be tall when they are twenty. But they may not be tall for their age at thirteen, because individuals differ in the age at which they undergo the growth spurt before puberty. In this case, the property of being taller than peers survives the big changes occurring at puberty. The same is true for many distinctive aspects of behaviour and personality. When these change permanently, as undoubtedly they sometimes do, it may not be because the person has passed through one of the supposed metamorphoses of development.

A change in control of an unchanged motor pattern might occur at a certain stage in development. Three-week-old kittens with their eyes recently open approach the mother from a distance for suckling but when they reach her, they search for a nipple with their eyes closed.[3]

---

3   Bateson, P. (2014), Behavioural development in the cat. In: Turner, D.C. & Bateson, P. (eds.), *The Domestic Cat: The Biology of its Behaviour*. Cambridge: Cambridge University Press, pp. 11–26.

Gradually, vision comes to trigger not only the approach to the mother but also the nipple-searching behaviour patterns that were formerly triggered by non-visual sensory systems. Such a change in control might easily be associated with a change in the rank order on the scale of measurement of the behaviour. In this case, the discontinuity can be investigated by an examination of the conditions that elicit the behaviour.

Every human individual must act many roles during his or her lifetime. Leaving aside the accuracy of how the roles are defined, how much of the individual's personality and distinctive behavioural characteristics fails to survive the crossing of the boundary between one of the distinct ages and the next? A big change occurs in humans between the ages of two and four with the emergence of language and an awareness of self. Few adults remember much of what happened to them in their first few years. Even if they are subjectively certain that they remember their birth, the corroboration is invariably suspect or missing. It might be argued that in the first few years children have no memories; nothing has been stored so nothing has to be erased. Such a view is clearly false. Young children have good functional long-term memories. In one experiment, for example, children around two years of age were asked to imitate actions that they had seen eight months before. They performed significantly better than children who had not previously seen these actions.[4] The absence of memories from infancy does not reflect an inability to form enduring memories at the time, suggesting instead that young children can remember things but that substantial reorganisation of memory occurs between the ages of two and four.

Mother cat suckling her kittens. Photo by Ashim 71 (2014), Wikimedia, https://commons.wikimedia.org/wiki/File:Mother_Cat_with_her_Kittens.jpg, CC-BY 4.0.

---

4   Strohl, K.P. & Thomas, A.J. (1997), Neonatal conditioning for adult respiratory behaviour. *Respir. Physiol.* 110.2–3, 269–275, https://doi.org/10.1016/s0034-5687(97)00092-3

Conclusions

The phenomena of discontinuities in development excite considerable interest because the changes may represent an alteration in the ecology of the developing individual. A variety of different explanations may be offered in any particular case. The phenomena are undoubtedly heterogeneous and so, while each explanation could apply some of the time, it is doubtful whether any one of them will apply all of the time. Some of the explanations point to deficiencies in method; some imply that after a loss of the original rank order on a particular measure, that order will be recovered, and some suggest that a change in rank order is permanent. The analogy with metamorphosis would be relevant only to phenomena brought about by the last group of processes. When behaviour at a given stage of development is matched to the ecological conditions in which the individual lives, the appearance of design is raised once again.

5. Early Experience and Later Behaviour[1]

The conviction that experience can exert a greater influence at some times of life than at others is deeply rooted in conventional thinking about humans and other animals. The disturbing picture of a child missing the developmental bus by not being treated in a particular way at specific times has been strongly challenged.[2] In many instances, it is possible to resolve the apparent contradiction between the view that the young are especially susceptible to particular experiences at particular times and the view that adults can also be affected by experience. This is because the effect of early experience involving influence or instruction from the environment can arise not so much through an incapacity to learn as through a reluctance to do so outside the period of sensitivity. When such reluctance can be overcome, it is possible for the older individual to learn about new things once again. It is important therefore to deal with the evidence for the effects of early experience separately from the mechanisms that control the timing of sensitive periods in development.

Many people who study the development of behaviour in humans feel uncomfortable, or even hostile, when evidence from animals is mentioned. It is as though the evidence is tainted with rigid determinism

---

[1] This chapter is drawn in part, with the publisher's permission, from Bateson, P. (1983), The interpretation of sensitive periods. In: A. Oliverio & M. Zappella (eds.), *The Behavior of Human Infants*. New York: Plenum Press, pp. 57–70, https://doi.org/10.1007/978-1-4613-3784-3. Much of this chapter is devoted to change and resistance to change in humans.

[2] Clarke, A.M. & Clarke, A.D.B. (1976), *Early Experience: Myth and Evidence*. London: Open Books.

and with the worst excesses of biologists intent on taking over the social sciences. It is a great pity that such attitudes are so prevalent because the phenomena encompassed by the evidence can provide highly promising gateways to an understanding of developing mental processes. A 'sensitive period', or one of its synonyms, ought merely to refer descriptively to the evidence that an individual's characteristics may be most strongly influenced by a given event at a certain stage of development. It is not an explanatory concept. It should not suggest that other events exert their maximum influence at the same stage. Nor should it carry implications about what might have given rise to the particular sensitive period that has been described.

In terms of an analogy with a moving train, the window of a particular compartment in the developmental train opens at a particular stage in the journey and then stays open. As a result of what an occupant learns about the outside world, it subsequently averts its gaze from anything strange. Because it can learn nothing until the window does open, the timing of the ending of the sensitive period is also dependent on the internal processes responsible for opening the window in the first place. If this analogy can be pressed a little further, it looks as though the occupant can, under certain circumstances, be persuaded to study strange things outside the train later in the journey, and when it does so it is influenced by what it sees.

It is possible to reconcile the view that early experience is important with the view that nothing is irreversible. The general point is that it may be possible for the distinctive features of behaviour to be formed in a particular stage in development and yet for the processes generating those features to be reactivated at much later stages in the life-cycle. An understanding of the mechanisms can explain why, under certain circumstances, the evidence for sensitive periods seems to evaporate.

Sigmund Freud's psychoanalytical theory, which was unusual at the time, reflected his belief that seemingly irreversible influences from childhood could be overcome in adults.[3] This view was central to Freud's method of therapy for those whose lives had been adversely affected by their early experience. Nowadays the idea is widely accepted and is implicit in the vast self-help industry, which is built on the supposition

---

3 Freud, S. (1905), *Drei Abhandlungen zur Sexualtheorie*. Leipzig: Franz Deuticke.

that people can change themselves. Indeed, the pendulum has swung so far that it often seems as though people should be able to change their behaviour and personality as readily as they change their clothes.

The older, and perhaps excessive, emphasis on early experience may have been rejected because of its implied pessimism that once someone had missed the developmental train, nothing could be done to help them thereafter. The grounds for optimism are in fact considerable, and evidence for sensitive periods early in development may be readily reconciled with evidence for subsequent changes in behaviour. This is most clearly seen when the experience that could cause the change is not normally encountered in later life. An unwillingness to eat novel food means that people will not encounter the flavours and textures that might change their preferences. But it is not just a matter of preference. The mechanisms in the brain that protect behaviour from change can be stripped away so that plasticity is once again possible.

Sigmund Freud. Photo by Max Halberstadt (1922), Wikimedia, https://commons.wikimedia.org/wiki/File:Sigmund_Freud_LIFE.jpg, Public Domain.

The behaviour patterns that are typical of gender, such as the style of play in boys and girls, may be amplified or minimised as the result of social influences from peers, teachers and parents. In the same way that a boy can become less 'boyish' in social circumstances that reduce gender differences, some shy children become less shy as they develop. Equally, some outgoing children become more withdrawn. The evidence that birth order has a significant effect on personality points once again to the subtle role of experience in development. Other factors, such as sudden changes in a family's economic circumstances, can also have big effects on what happens to a child.

Washing the brain

Academics are sometimes caricatured (not entirely unfairly) as accumulating more and more detailed knowledge about a subject on which their focus becomes ever narrower. In the image of Waddington's epigenetic landscape (see Chapter 3), they have descended into an intellectual valley from which escape becomes increasingly difficult. They are stuck in their own silos of knowledge. Even so, some scholars manifestly do break out of these narrow confines. Indeed, a much admired feature of the wisest academics is their ability to make connections between different bodies of knowledge. Is such willingness to branch out equally true for their more deeply seated beliefs and attitudes, such as their political persuasion or their sociability? Most people would say 'no'. Values are established in early life and, it is supposed, remain firmly fixed thereafter.

American public opinion which had been comfortable in this belief took a jolt in the Korean War. About a third of the 7,000 American prisoners of war collaborated with their Chinese and North Korean captors, and 21 refused to return to the USA when the war was over. These 'conversions' generated consternation in the USA and stimulated an intense examination of the techniques used by the captors of the prisoners of war. Many of the apparent conversions turned out to have been little more than the effects of prolonged deprivation of sleep or, in some cases, self-preserving attempts to secure better living conditions. The so-called brainwashing methods were neither subtle nor sophisticated. Even so, some of the prisoners who had been subjected to terror, physical hardship and intensive indoctrination did seem to have changed their values and political allegiances in a more fundamental way.

The origins of brainwashing lie much further back than the Korean War. Echoes can be found, for instance, in the Christian revivalist conversions in eighteenth-century America. During a religious crusade in Massachusetts in the 1730s, the theologian Jonathan Edwards discovered that he could make his 'sinners' break down and submit completely to his will. He achieved this by threatening them with hell and thereby inducing acute fear, apprehension and guilt. Edwards, like many other preachers before and after him, whipped up the emotions of his congregation to a fever-pitch of anger, fear, excitement and nervous tension, before exposing them to the new ideas and beliefs he wanted them to absorb. To this day, live rattlesnakes are passed around some congregations in the southern parts of the USA; the fear they induce can impair judgement and make the candidates for conversion more suggestible. Once this state of mental plasticity has been created, the preacher starts to replace their existing patterns of thought. And constant fear is, of course, a hallmark of totalitarian regimes where dissenting individuals live under the unremitting threat of detention, torture or execution.

The British psychiatrist William Sargant, working in the middle of the twentieth century, was deeply interested in the mind-moulding techniques, and noticed the importance of high emotion in the process of religious conversion.[4] He drew on a wide range of human experience, including that of military brainwashing. He extended his inquiry to the beneficial uses of stress in psychotherapy. In the Second World War Sargant tried to help soldiers suffering from battle fatigue. As part of the therapy, he and his colleagues would deliberately arouse strong emotions in their patients, about events that had no direct connection with the trauma they had experienced.

Sargant argued for the importance of the emotional reaction in therapy, whereby patients are made anxious, guilty and even angry by their therapist and, in consequence, become able to change their previous patterns of behaviour. Sargant's method has something in common with the psychological therapeutic technique of 'flooding', in which someone suffering from a phobia is deliberately frightened in the presence of the object or situation towards which they are phobic — for instance, by placing a large spider onto the chest of the patient who is terrified of spiders.[5] Contrary to what intuition might suggest, the patient's phobia may sometimes be reduced.

It is common practice around the world for army recruits to be treated brutally in the early stages of their training. The individual is broken down through physical and mental pressure before being rebuilt

---

4   Sargant, W. (1957), *Battle for the Mind: A Physiology of Conversion and Brainwashing*. London: Heinemann.

5   References to flooding in psychiatry are given in https://en.wikipedia.org/wiki/Flooding_(psychology)

in the form required by the military. The recruits are verbally abused, made to perform pointless menial tasks, and forced onto long marches carrying heavy equipment. By degrees their platoon becomes their family. Away from armies, other methods for inducing psychological plasticity include social isolation, fasting, lowering blood glucose with insulin, physical discomfort, chronic fatigue and the use of disturbing lighting and sound effects.

The so-called Stockholm syndrome, also known as 'terror bonding' or 'trauma bonding', may be yet another instance of psychological plasticity induced by emotional trauma. The term takes its name from an incident in Stockholm in the 1970s, in which a woman who was taken hostage in a Stockholm bank following an unsuccessful robbery formed a strong and long-lasting emotional bond with her captor. She even remained faithful to him during his subsequent imprisonment. Her strange reaction was not unique. Many other victims of violent hostage-taking have ended up siding with their captors against the authorities who were trying to rescue them. Being taken hostage is obviously a traumatic experience, and the hostage-takers may be equally frightened because their lives are on the line as well. In such circumstances, where hostage and captor are exposed to each other while both are emotionally highly aroused, a strong emotional bond may form, bizarrely uniting them against the world outside. As with the various military, political, religious and therapeutic techniques for changing the way adults think and behave, the crucial element is the combination of psychological stress and suggestion.

Comparable cases in which trauma has induced behavioural plasticity have been observed in other species as well. Adult wild horses are commonly 'broken' by traumatising them whilst exposing them to humans. A traditional but brutal method involves near-strangulation with a rope; even the wildest of wild horses can be reduced to gentle submissiveness in as little as 15 minutes using this technique. Unsocialised adult dogs can similarly be induced to form strong attachments to humans by means of traumatic discipline. (The fact that these practices work does not make them desirable.) An anecdotal but nonetheless illuminating case concerned a remarkable change in an adult female Soay sheep, which was part of a small flock living in the grounds of the University Sub-Department of Animal Behaviour at Madingley, near Cambridge. The Soay sheep were wild, avoiding human beings, and the female in question was no exception. Then, one spring, she had a particularly difficult time giving birth. It was eventually necessary for Sub-Department's staff to assist, by catching the mother and pulling her lamb out. This was undoubtedly a traumatic experience for her. Ever afterwards, until she died, this sheep remained strongly attached to humans and would follow people around as they moved about the grounds of the laboratory. The trauma of the birth, combined with simultaneous exposure to people, brought about a profound and long-lasting change in this animal's behaviour.

Neurobiology

The concept of extreme fear or emotional arousal inducing plasticity helps make sense of many diverse examples of behavioural change. What might be the neurobiological mechanisms underlying this effect? How does trauma make someone susceptible to fundamental changes in their thoughts and values? What might be the biological link between psychological stress and the processes of plasticity and change in the nervous system?

High levels of psychological stress are associated, amongst other things, with the rapid synthesis and turnover of the neurotransmitter substance noradrenaline. This chemical messenger of the nervous system has been implicated as an enabling factor in making the adult brain become plastic again. Noradrenaline (known in the USA as norepinephrine) is released in the mammalian brain, at the endings of neurons throughout the body, and from the adrenal glands just above the kidneys. It is released, amongst other things, in response to psychological stress; in humans, a mildly stressful situation such as giving a public speech will typically elicit a 50% rise in the amount of noradrenaline circulating in the bloodstream.

An experiment on the visual system of cats gave some valuable insights into the connection between noradrenaline and plasticity.[6] The mammalian visual system is normally changeable only during an

---

6   Pettigrew, J.D. & Kasamutsu, T. (1978), Local perfusion of nor adrenaline maintains visual cortical plasticity. *Nature* 271: 761–763, https://doi.org/10.1038/271761a0

early stage in the individual's life. The capacity of an eye to stimulate neurons in the visual cortex of the cat's brain depends on whether that eye received visual input between about one month and three months after birth. If one eye is deprived of visual stimuli during this period it virtually loses its capacity to excite cortical neurons thereafter, no matter how much visual stimulation it receives. The eye consequently becomes functionally blind, even though it remains physically unimpaired. Once the dominance of the other eye is established, it is exceedingly difficult to change the relationship with the unused eye. Similarly, binocular vision cannot easily be disrupted in normally reared individuals once it has become established. Infusing noradrenaline into one hemisphere of the visual cortex of older cats can re-establish plasticity and enable further change to occur in response to visual experience. If normally-reared animals are deprived of the use of one eye during the period of noradrenaline infusion, binocular control of the neurons is lost in the visual cortex of the hemisphere that was infused. No such change occurs in the visual cortex of the other hemisphere. In other words noradrenaline can reverse in adulthood what would otherwise be unchangeable.[7]

If one eye is occluded in early life by, say, infection, the other eye becomes dominant for the rest of the cat's life. Photo by Patrick Bateson, CC-BY 4.0.

Continuity and change

The variety and complexity of behaviour and its underlying psychological systems inevitably means that any sweeping statement about the possibility of change must eventually come unstuck. The self-help industries that promise relief from shyness, depression, sloth, obesity, or addiction to nicotine deliver results only some of the time. Once developed, some patterns of behaviour are strongly buffered against subsequent change. Preference for certain types of food and

---

7   Baroncelli, L. *et al.* (2016), Experience affects critical period plasticity in the visual cortex through an epigenetic regulation of histone post-translational modifications. *J. Neurosci.* http://dx.doi.org/10.1523/JNEUROSCI.1787-15.2016

for particular places tend not to change. They may be stable for good reasons, since change can be disruptive and costly. On the other hand, not to change may, in certain circumstances, carry an even bigger cost, which perhaps explains why behavioural characteristics tend to become plastic under conditions of stress.

When one aspect of behaviour changes it does not imply that everything else must change as well. But whatever the complexities of development and the inadequacies of current understanding, it is clear that adults really *are* capable of changing — more so, perhaps, than many suppose.

The psychiatric evidence for rehabilitation leaves untouched the question of which forms of behaviour, once developed, are most strongly buffered against subsequent change. It seems likely that in adults cognitive processes are more easily changed than those underlying their emotions.[8] Adaptations that protect certain well-developed preferences and habits from alteration are to be expected. Fear of novelty, though general in its effects, would serve precisely this function.

Conclusions

A positive point that emerges is that evidence for sensitive periods in development can be readily reconciled with evidence for subsequent changes in behaviour. This is most clearly the case when the form of treatment involves experiences that would not normally be encountered in later life. Once the mechanisms protecting behaviour from change are stripped away by suitable treatment, change resulting from renewed plasticity is once again possible. Any changes in sensitivity that are found imply no particular mechanism. The search for what might generate a sensitive period in development is a separate enterprise. Even so, the stability of some aspects of behaviour, once formed, can be changed under certain conditions such as chronic stress. The flexibility make good sense in biological terms since it enables the individual to cope in a changed environment. The capacity has the appearance of good design.

---

8   This conclusion is supported by the work on Cognitive Behavioural Therapy which is used to treat a number of psychiatric disorders (e.g. McKay D. *et al.* (2015), Efficacy of cognitive-behavioral therapy for obsessive-compulsive disorder. *Psychiatry Research* 225.3, 236–246, http://dx.doi.org/10.1016/j.psychres.2014.11.058).

6. Communication between Parents and Offspring[1]

How offspring respond to the behaviour of their parents raises important issues about their development. In studies of animal communication it has sometimes been argued that all the activities directed by one individual towards another are manipulative.[2] Sometimes this view is clearly correct. Its most obvious form is in the interactions between species when one species manages to control the behaviour of another as if it were a puppet. A striking example is the European cuckoo. The mother cuckoo lays each egg in the nest of another species such as the reed warbler. The egg very closely resembles the egg of the host. The young cuckoo hatches before the reed warblers' chicks and ejects the competition from the nest. Then the young cuckoo successfully persuades the unfortunate warbler foster parents to feed it, even when it is twice their size. By looking like a super-offspring, the cuckoo successfully exploits the normal pattern of interaction that exists between parent and young.

The essentially competitive process of Darwinian evolution does not necessarily imply a competitive outcome. On the contrary, a great deal of communication involves signals that carry real information and cooperation in which all the participants benefit by working with each

---

1 Part of this chapter is taken, with permission, from Bateson, P. (1990), Animal communication. In: *Ways of Communicating*, ed. by D.H. Mellor. Cambridge: Cambridge University Press, pp. 35–55.

2 Dawkins, R. & Krebs, J.R. (1979), Arms races between and within species. *Proc. R. Soc. Lond B*, 205.1161, 489–511, https://doi.org/10.1098/rspb.1979.0081

other. One explanation for cooperation is that, at least in the past, the aided individuals were relatives; cooperation is like parental care and has evolved for similar reasons. Another is that co-operating individuals jointly benefited even though they were not related; the co-operative behaviour has evolved because those who engage in it were more likely to survive as individuals and reproduce than those that did not. Once again the force of this particular argument can be seen most clearly in communication between different species.

Some small fish, which are conspicuously marked but highly suitable as a morsel of food, clean the teeth of big predator fish. Before they do their job, the cleaner fish perform a characteristic waggling swim in front of the monster. This inhibits the normal feeding response and the great predator opens its mouth, allowing the little fish in. When the big fish needs to eat other little fish, it signals it is switching back into normal hunting mode by jerking its jaw in a particular way. The little cleaner fish scuttle for cover and the mutually beneficial symbiotic arrangement is preserved.

Small cleaner fish servicing a big predator fish. Photo by Richard Ling (2005), Wikimedia, https://commons.wikimedia.org/wiki/File:Epinephelus_tukula_is_cleaned_by_two_Labroides_dimidiatus.jpg

In co-operating animals of the same species, the mutual benefits of working together can be greatly enhanced if information about the state of the external world is transmitted from one individual to another. One of the most extraordinary and well-analysed examples of such transmission is the so-called dance language of honey bees.[3] The characteristics of the waggle dance performed in the hive provide crucial information about where the returning bee successfully foraged. The duration of the dance circuit is strongly correlated with the distance from the hive to the food. In a darkened hive the angle of orientation of the central segment

---

3   Munz, T. (2016), *The Dancing Bees: Karl von Frish and the Discovery of Honeybee Language*. Chicago: University of Chicago Press, https://doi.org/10.7208/chicago/9780226021058.001.0001. A video of the honey bees' waggle dance is available at https://www.youtube.com/watch?v=-7ijI-g4jHg

of the dance with respect to the vertical is related to the angle between the food source and the sun's position.

The most interesting test-bed of evolutionary thinking about communication is provided by those situations in which individuals compete with each other over resources. Advocates of the manipulation view urge that a trustworthy mode of communication is always open to cheating in which the cheater exploits information provided by its opponent and gives nothing away about itself. As a consequence, it is much more likely to win the resource and likely to reproduce faster. Before long, it is argued, cheating will have evolved to become the dominant mode of behaviour.

This argument sounds convincing, but most people who know a particular species well quickly develop a good intuitive sense of whether an animal is likely to attack or escape by observing its body and facial postures. After the initial location of the opponent, several levels of escalation may precede an actual fight, involving easily recognised and increasingly energetic displays. Neither party will benefit from getting hurt and, in the majority of cases, the disputes are settled without serious damage on either side. The encounter can break off at any stage during the process of escalation. The contests are usually won by the larger individual, but if the opponents are of equal size, they are usually won by the holder of the resource. The behaviour of each individual at each stage of escalation indicates how serious it is about continuing.

From a large number of quantitative studies, it is clear that escape *is* well predicted by certain patterns of behaviour. If one animal suddenly turns tail, it is liable to be attacked and might get injured. The advantage to the loser of not being misunderstood and expressing the animal equivalent of a white flag is obvious. The benefit to the winner from responding appropriately to such a signal is that it does not risk injury by escalating the conflict into a real fight unnecessarily. The argument is, therefore, that a form of behaviour which effectively negotiates the end of a conflict can be evolutionarily stable.

While most biologists placed great emphasis on dissimulation, one took a very different line. The Israeli biologist Amotz Zahavi argued that signalling that carried a handicap to the signaller was honest.[4]

---

4 Zahavi, A. & Zahavi, A. (1997), *The Handicap Principle: A Missing Piece of Darwin's Puzzle*. Oxford: Oxford University Press.

Zahavi was treated as an eccentric since the initial attempts to model his ideas about handicaps failed formally to substantiate them. More recently, theoreticians have been able to satisfy nearly everybody else that if a signal carries a cost, it may also be reliable in the sense of carrying accurate information.[5] Honest signalling is now widely accepted, although arguments continue about whether honest signals must carry some cost.

A crucial question is where the balance is struck between signalling real information about internal state and signalling misinformation. Consider analogies with the human game of poker. On the one hand, if players can get away with bluff, they will make money. On the other, if players bluff against opponents who have really good hands, they may end up very much worse off than if they had decided to throw in their bad hand before they had raised the bet too far. So something equivalent to negotiation might be expected. For each individual the optimal outcome of such negotiation should represent a balance between the costs of escalating the conflict to likely damage to the pocket (or in case of fights, the body) and the benefits of winning the resource easily. An individual that escalates without assessment is in danger of finding itself in a fight with a much stronger opponent.

Red deer stag roaring at competitors in rutting season. Photo by Bill Ebbesen (2009), Wikimedia, https://commons.wikimedia.org/wiki/File:Red_deer_2009.jpg, CC BY 3.0.

Evolutionary pressure is present for some exaggeration of fighting ability and many species puff themselves up in various ways, making themselves look more fearsome than they really are. Even so, honest advertisement of strength, providing cues that cannot be faked, may count most in the long run. One example may be the male red deer, which competes vigorously with other males for opportunities to mate. Fights occur, of course, but conflicts are most often settled by bouts of roaring at each other.

---

5   Grafen, A. (1990), Biological signals as handicaps. *J. theor. Biol.* 144.4, 517–546, https://doi.org/10.1016/s0022-5193(05)80088-8

The roaring is extremely exhausting for the animals and the one that can keep it up for longest is also likely to be the stronger. In the roaring contests, both individuals increase the rate of roaring until one seems to recognise it is outclassed and retreats. Recordings of red deer roars were played to a real stag.[6] Unlike the stag, the tape-recorder did not get tired and when the tape-recorder roared at a high rate, the stag roared less. It seems as though the stag has been forced into accepting that it was dealing with a much more powerful opponent. While fake characteristics may gain short-term successes, those individuals that ignore such cues and focus on reliable sources of evidence about their opponents may eventually do better in the course of evolution.

Parents and offspring

Evolutionary theories about parent-offspring relationships have undergone a similar transformation to those about communication. The prediction used to be that young will demand food and care from their parents at weaning whereas the parents' interests are best served by reserving their efforts for future offspring. The American biologist Robert Trivers revolutionised the study of parent-offspring relationships when he pointed out that the long-term interests of the parent are not identical with those of its offspring.[7] A parent may increase its reproductive success by weaning its young earlier than is best for them, because it saves itself from becoming exhausted and is thus able to have a larger number of offspring than would otherwise have been the case. As a consequence, aggression between parents and offspring is to be expected, particularly at the time of weaning, and the communication between them was not thought to be reliable.

Trivers's stimulating contribution has been responsible for a large theoretical literature. The insights derived from evolutionary theory have been questioned partly because of changes in theoretical stance about the benefits of reliable signals. More seriously, perhaps, the

---

6   Clutton-Brock, T.H. & Albon, S.D. (1979), The roaring of red deer and the evolution of honest advertisement. *Behaviour* 69.3, 135–144, https://doi.org/10.1163/156853979x00449. A video of a male deer's roaring is available at https://www.youtube.com/watch?v=76pRLpwumjc

7   Trivers, R.L. (1985), *Social Evolution*. Menlo Park, CA: The Benjamin/Cummings Publishing Company.

evolutionary argument had led to an unjustified expectation that conflict of evolutionary interest necessarily implies behavioural conflict. Behavioural aggression between parent and offspring should be referred to as 'squabbling', thus avoiding the ambiguous use of 'conflict'. Even so, the behavioural evidence does not fit the theory especially well.

Two-year-old tantrums in humans have been interpreted in precisely the opposite sense to the theory of conflicts. When they occur, tantrums seem to be concerned with the difficult process of establishing some autonomy from the parent. Well-attached children with sensitive mothers do not seem to have tantrums, using their parents as a secure base from which to explore the world. Rhesus monkey mothers may actively reject their offspring around the time of weaning, while their offspring play an increasingly prominent role in initiating contact with their mother as they grow older.[8] Squabbles did not arise in this species as a result of mothers' reduction in suckling when they conceived their next offspring. Nor did mothers that failed to conceive have more demanding infants, as would have been predicted from the theory of parent-offspring conflict.

Occasionally, weaning squabbles between mother and young do occur, particularly if the mother is already pregnant again. In field studies of baboons, conflicts of interest did not invariably lead to such squabbles. Evidence from a variety of mammals suggests that maternal aggression often does not occur at all and, if it does, it is seen at quite different stages in development from that in which the process of weaning occurs.[9]

Perhaps the most convincing evidence against the early manipulative predictions derived from parent-offspring conflict theory was that mothers both monitor and respond to the progress of their current offspring. Mothers may actively compensate for the slow development of their young or delay conception of the subsequent offspring if the present one is progressing slowly. Rat mothers continue to lactate for longer than would otherwise have been the case when they are given

---

[8] Gomendio, M. (1991), Parent/offspring conflict and maternal investment in rhesus macaques. *Anim. Behav.* 42.6, 993–1005, https://doi.org/10.1016/s0003-3472(05)80152-6

[9] Bateson, P. (1994), The dynamics of parent-offspring relationships in mammals. *Trends Ecol. Evol.* 9.10, 399–403, https://doi.org/10.1016/0169-5347(94)90066-3

pups younger than the ones they had been suckling, again suggesting that the mother is not necessarily driving the weaning process without regard for her offspring's condition.

From the offspring's standpoint, young laboratory rats seem to wean themselves spontaneously; a stage is reached in development when they actively choose solid food in preference to milk. When the offspring's metabolic costs go up as they get older and larger, they may be forced to wean themselves because they simply do not get enough energy from their mother's milk. In domestic cats, the rate of growth starts to slow down around 20 days after birth, particularly in kittens of nutritionally stressed mothers, and then suddenly speeds up again around 30 days as they start to process solid food efficiently. The discontinuity occurs at different ages in different animals and is less apparent, as a consequence in averaged data. The slowing down of growth before the discontinuity is much more pronounced in larger litters that impose greater energetic demands on their mothers. The eventual inadequacy of the mother to provide for all the nutritional needs of the young is particularly evident in fur seals in which the mother disappears for several days while she stokes up for the next bout of lactation. Eventually what she can provide is less than what her pup needs.[10]

The empirical study of behaviour has forced a reappraisal of the optimal route to maximum reproductive success in mothers and the optimal route to highest probability of survival in their offspring. The evolutionary arguments needed to be looked at again, not simply from the standpoint of honest signalling but also bearing in mind how an individual's state might affect the optimal weaning time from both the standpoint of the mother and that of her offspring.[11] In species that can breed more than once, the mother has a reasonable chance of successfully raising more than one offspring. The current offspring will only survive to independence after a certain minimum amount of care has been given to it. Even when the mother is well fed, the longer she cares for the current offspring, the worse her state may become, given the substantial energy costs of lactation. As a consequence, the costs in terms of the diminished probability of rearing more young may soon outweigh the marginal benefits of giving more milk to the current young.

---

10   Ibid.
11   Ibid.

The probability of having another offspring ought to influence the time and energy devoted to the current one. For example, the mother may have to face a winter in which she cannot breed and which she might not survive. Furthermore, matters may be made worse if finding a mate takes time. With a relatively low probability of breeding again, her lifetime reproductive success may be maximised if she gives higher priority to the current offspring than to those that are as yet unborn. Conversely, if she has mated immediately after the birth of the current young and is pregnant while she lactates, as is possible in species such as the cat and the rat, she may give lower priority to the offspring she is currently looking after than if she were not pregnant. Under most environmental conditions, the optimum weaning time should be earlier than for non-pregnant mothers.

The reactions of an offspring to its mother's conditional responses to the environment should also be conditional. In many species, the relation between the effect of its behaviour on its probability of surviving before weaning and its probability of surviving after weaning must both be taken into account. This is because offspring survival does not simply depend on getting as much milk as possible from the mother while she is able to provide it. Other benefits, such as the protection the offspring might have derived from her in the post-weaning period, would be lost if it were so demanding in the pre-weaning period that it seriously damaged her health. In the early stages of lactation the offspring's survival depends utterly on receiving immediate maternal attention. Moreover, since meeting its needs is relatively easy for the mother because of the offspring's low body mass compared to that of the mother, the longer-term consequences for post-weaning care of being demanding are not heavy. As the offspring grows, these delayed effects of being demanding start to rise and, at the same time, the chances of the young surviving without maternal milk begin to improve. By the time the offspring has reached the late pre-weaning stage, its peak demands have dropped to low values.

The optimal weaning ages for offspring depend greatly on how much care is given after weaning. If little is characteristically given by the species, optimal demand by the offspring remains high until the demands of switching to primarily solid food become paramount. The greater the amount of post-weaning care the mother gives, the earlier the age relative to weaning when her offspring's demanding behaviour should start to decline. The general point is that an offspring benefits from reducing its demanding behaviour as it gets older and also benefits from monitoring maternal availability which is a measure of her state.

Each mother is sensitive to the condition of its offspring so that, if it is still in need of care but reasonably well developed, she may forego a breeding opportunity in order to nurture it through to independence. Young are sensitive to the condition of their mothers and adjust their pattern of development accordingly, since their mothers respond not only to the state of the young but also to their own condition.

The offspring may also need to prepare for the probable world in which they have to grow up. Members of the same species, the same sex and the same age sometimes differ dramatically from one another. 'Alternative tactics' within a species commonly arise because an individual has the capacity to respond in more than one way according to environmental conditions or its own body state. Such conditional responses during development are well known in the social insects, in which one sister might be adapted for producing thousand of eggs, another has massive jaws used in defence of the nest and another is equipped with foraging skills never expressed by the other two. Similar environmentally-induced differences occur frequently in mammals. Young mammals may pick up crucial information from their mothers about when to wean themselves and how to develop later on the basis of cues that they pick up from their nursing mothers.

The process of weaning in mammals is not readily explained in terms of conflicts of interest that inevitably lead to squabbling. From a behavioural standpoint, the interplay between parent and offspring is dynamic. The mother has to balance the maintenance of her body condition, easily debilitated by energetically costly lactation, against responding sensitively to the needs of current offspring. If the current offspring lag behind in development, the mother may increase her reproductive success extending her parental care.

For its part, each offspring balances demanding maximum resources and attention from its mother up to the last moment she is prepared to feed it against being sensitive to its mother's state. The young have to pay attention to the condition of the mother because of the need to take into account both the immediate effects of maternal care on survival but also the post-weaning contributions of the mother. Moreover, to adopt characteristics that will be appropriate to the environment into which they will have fend for themselves, the young may need to respond to information provided by the mother. The mother and her offspring are unlikely to have perfect knowledge of each other either initially or during development. Therefore each needs to monitor its own and the other's often rapidly changing state.

Conclusions

Communication is not simply about manipulating another individual. The emerging picture is more nuanced than that. Sensitivity to the condition of the receiver of the signal is often adaptive to the sender. So the sender benefits in Darwinian terms and the competitive evolutionary process produces behaviour that is apparently more cooperative than has sometimes been supposed. In short, the mother-offspring relationship is a system that is apparently well-designed for both parties.

7. Avoiding Inbreeding and Incest[1]

Finding a compatible partner is an important part of reproductive behaviour in many animals in which mates are chosen carefully. Members of different species do not make good mates. At the other pole, too much inbreeding can also reduce reproductive success.

Inbred animals are more likely carry some damaging genes. Most potentially harmful genes are recessive and are therefore harmless when they are paired with a dissimilar gene become damaging in their effects when combined with an identical gene. They are more likely to be paired with an identical recessive gene as a result of inbreeding. The genetic costs of inbreeding arising from the expression of damaging recessive genes are the ones that people usually worry about.[2] Recessive genes are less of a problem in mammals than they are in birds because mammals generally move around less and they may live in quite highly inbred groups where the harmful alleles have been purged. The most

---

1   This chapter is drawn, with permission from the publishers, from: Bateson, P. (1983), Optimal outbreeding. In: P. Bateson (ed.), *Mate Choice*. Cambridge: Cambridge University Press, pp. 257–277 and Bateson, P. (2004), Inbreeding avoidance and incest taboos. In: Wolf, A.P. & Durham, W.H. (eds.), *Inbreeding, Incest and the Incest Taboo*. Stanford: Stanford University Press, pp. 24–37.

2   Across Asia, the effects of genetic disorders are becoming increasingly obvious. This change is especially important among the children of couples who have married cousins and is also found in migrant communities resident in North America, Western Europe and Australasia who continue the tradition of close kin marriages (Bittles, A.H. (2003), Consanguineous marriage and childhood health. *Developmental Medicine* 45.8, 571–576, https://doi.org/10.1111/j.1469-8749.2003.tb00959.x). The unfortunate consequence arises from practicing first cousin marriages generation after generation. In relatively inbred animal populations the preferred genetic distance between mates may be greater than in more outbred populations thereby offsetting the effects of mating with close kin (see Bateson, P. (1983), *Mate Choice*).

important biological cost of excessive inbreeding is that it negates the benefits of the genetic variation generated by sexual reproduction. If an animal inbreeds too much, it might as well make copies of itself without the effort and trouble of courtship and mating.

On the other side, excessive outbreeding also has costs. For a start, excessive outbreeding disrupts the relation between parts of the body that need to be well adapted to each other. The point is illustrated by human teeth and jaws. The size and shape of teeth are strongly inherited characteristics. So too are jaw size and shape, as may be seen in many paintings of the Hapsburg family, scattered in the museums around the world. The Dürer painting of the Holy Roman Emperor Maximilian I reveals the large Hapsburg jaw, which was even more pronounced in his highly inbred great-great-great-grandson, Philip IV of Spain, shown in the painting by Velasquez. The potential problem arising from too much outbreeding is that the inheritance of teeth and jaw sizes are not correlated. A woman with small jaws and small teeth who had a child by a man with big jaws and big teeth lays down trouble for her grandchildren, some of whom may inherit small jaws and big teeth. In a world without dentists, ill-fitting teeth were probably a serious cause of mortality. This example of mismatching, which is one of many that may arise in the complex integration of the body, simply illustrates the more general cost of too much outbreeding.

Two members of the Hapsburg family separated by five generations. Albrecht Dürer, Portrait of Maximilian I, Holy Roman Emperor (1519), Kunsthistorisches Museum, Vienna. Wikimedia, https://commons.wikimedia.org/wiki/File:Albrecht_Dürer_-_Portrait_of_Maximilian_I_-_Google_Art_Project.jpg, Public Domain. Diego Velázquez, Portrait of Philip IV (1656), National Gallery, London. Wikimedia, https://commons.wikimedia.org/wiki/File:Philip_IV_of_Spain.jpg, Public Domain.

With costs accruing to both inbreeding and outbreeding, Darwinian evolution is presumed to have operated on mechanisms involved in mate choice to minimise both. The outcome of Darwinian evolution is a preference for a mate that is not too closely related and not too distantly related.[3] Hybrid vigour is so dramatic when it occurs that it seems to argue against the view that marginal outbreeding is beneficial. Such hybrids are usually infertile and the original parents will have few or no grand-offspring. This means that tests of outbreeding depression should be made in the natural environment where the benefits of co-adapted gene complexes can be revealed.

In experiments with the mountain delphinium the largest number of seedlings was produced by crosses between plants that were ten metres apart. Plants that were self pollinated and those that were crossed with plants 1000 metres away, both gave rise to significantly smaller numbers of seedlings.[4]

This delphinium produces most seeds when crossed with plants 10 metres away from it. *Delphinium nuttallianum*. Photo by Walter Siegmund (2009), Wikipedia, https://commons.wikimedia.org/wiki/File:Delphinium_nuttallianum_9179.JPG, CC-BY 3.0.

In animals direct evidence for the genetic costs of outbreeding is still relatively slender, although the examples are starting to multiply. Some of the studies of humans suggest that fecundity is related to the similarity between spouses. For example, in one study the more alike human couples were on 17 out of 19 measures of the body (such as forearm length, height and ear length), the more children they had. Although most correlations were positive, each

---

3   The idea of balance between inbreeding and outbreeding was first suggested by Wright, S. (1933), The roles of mutations, inbreeding, crossbreeding and selection in evolution. *Proc. VIth, Internat. Congr. Genetics*, vol. 1, 356–366, available at http://www.esp.org/books/6th-congress/facsimile/contents/6th-cong-p356-wright.pdf

4   Price, M.V. & Waser, N.M. (1979), Pollen dispersal and optimal outcrossing in *Delphinium nelsonii*. *Nature*, 277.5694, 294–297. Independently of the use of 'optimal outbreeeding' (see Bateson, P. (1978), Sexual imprinting and optimal outbreeding. *Nature*, 273.5664, 659–660, https://doi.org/10.1038/273659a0) they used the term 'optimal outcrossing'.

correlation coefficient was low and could, of course, be explained by shared associations between the measures and a third variable such as social class. In a study of Icelandic partners, those who were distant cousins had more grandchildren than those who were more closely or more distantly related. The best outcome in terms of biological adaptiveness is to have the most grandchildren when other things such as social status are equal.[5]

Not all the costs of outbreeding too much are genetic in character. If and when they operate, these non-genetic costs further complicate the interpretation of laboratory breeding experiments aimed at settling whether or not outbreeding can be costly. For instance, two unrelated individuals in the laboratory may produce many offspring and, because they share adjoining cages and therefore common antibodies, they incur no cost from being exposed to pathogens carried by the other. It could be a different story in the natural environment. Similarly, the advantage of using skills acquired for dealing with the local environment could counteract the genetic advantage of moving into another area prior to breeding. Yet the non-genetic advantage could not be assessed in a laboratory experiment. Nor could the various costs when moving away from the natal area such as increased risks from predation. Some caution is needed before jumping to conclusions about just where the balance between inbreeding and outbreeding is likely to be struck.

If an animal does best by choosing a mate that is neither a close relative nor totally unrelated to it, what mechanisms could it use? Of the various types of explanation that have been offered, two are likely to be important. The first proposes that prior to mating, members of one sex move away from the area where they were hatched or born.[6] Providing they do not move too far, their mates are likely to bear some genetic relationship to themselves and so optimal outbreeding could be achieved. The second possibility is that animals are able to recognise close kin and, on the assumption that physical appearance is a measure of genotypic similarity, choosing a mate that looks, sounds or smells a

---

5   Helgasson, A. *et al.* (2008), An association between the kinship and fertility of human couples. *Science*, 319.5864, 813–816, https://doi.org/10.1126/science.1150232

6   Greenwood, P.J. (1980), Mating systems, philopatry and dispersal in birds and mammals. *Anim. Behav.* 28.4, 1140–1162, https://doi.org/10.1016/s0003-3472(80)80103-5

bit different but not too different from close kin will result in optimal outbreeding. These two explanations are not mutually exclusive. Some species could employ both mechanisms. The evidence suggests that both are found in the animal kingdom.

First, in many species of bird and mammal one sex moves out of the natal area prior to breeding. In most species of birds, females move away although exceptions exist such as the snow goose in which the male is unlikely to return to the natal area. In most species of mammal the males are more likely to move away but here again exceptions are known such as the chimpanzee in which the females move away from their natal group. The costs of travel can be quite considerable and usually the distance travelled by the sex that moves is not great. The net effect of restricted movements in one direction and returns by offspring in the next generation could be an overall population that was quite highly inbred. An important question remains whether such a system would be sufficiently finely tuned to preserve the optimal balance between inbreeding and outbreeding.

Recognition of kin, the other suggested mechanism for optimal outbreeding, could be accomplished in one or two ways. One possibility is that the genes that influence an animal's external appearance also directly influence its ability to recognise another animal very much like itself without the involvement of any learning process. More plausibly, the animal learns the characteristics of close kin, or failing that of itself, and can then recognise novel individuals that are similar to kin. In Japanese quail a first-cousin is preferred by both sexes over siblings and also over unrelated birds.[7]

Natural experiments have been performed unwittingly on humans. The most comprehensive evidence has come from the marriage statistics from Taiwan in the late nineteenth and early twentieth century, when Taiwan was under Japanese control. The Japanese kept detailed records for the births, marriages and deaths of everyone on the island. As in many other parts of South East Asia, marriages were arranged, and they occurred mainly and most interestingly in two forms. The 'major' type of marriage was the conventional one in which the partners first met each other when adolescent. In the 'minor' type of marriage, the wife-to-be

---

[7] Bateson, P. (1982), Preferences for cousins in Japanese quail. *Nature* 295.5846, 236–237, https://doi.org/10.1038/295236a0

was adopted as a young girl into the family of her future husband. In minor-type marriages, therefore, the partners grew up together like siblings. Later in life their sexual interest in their partner was assessed in terms of divorce, marital fidelity and the number of children produced. By all these measures, the minor marriages were conspicuously less successful than the major marriages.[8] Typically, the young couples who had grown up together from an early age, like brother and sister, were not much interested in each other sexually when the time came for their marriage to be consummated. Girls who were adopted into families before the age of three and then married their adopted 'brother' had a lower fertility than girls adopted later.

In the past Israeli kibbutzniks who grew up together like siblings rarely married each other. The few who chose to marry within their peer group were usually those who had entered the kibbutz after the age of six and therefore had not grown up with their future spouses.

Early experience and sexual attraction

Neither of the evidence from Taiwan or Israel means that the learning process that affects adult sexual preferences is completed early in life. If children grow up together and consequently see a lot of each other, they revise the ways in which they recognise each other; this goes on until they are sexually mature. By the time they are three, children are highly conscious of their own sex and are much less likely to play with somebody of the opposite sex, particularly a child who is not well known to them.[9] It seems plausible then that a girl who is adopted when over three will be viewed as a stranger by the boy, and treated differently from a girl who is adopted when younger.

How could finely tuned sexual preferences arise from early experience? The responsiveness to the familiar could be reduced by mere exposure, and consequently individuals that differed slightly from the known standards would be most attractive. A simple way of producing a finely tuned preference displaced away from the familiar

---

[8] Wolf, A.P. (1994), *Sexual Attraction and Childhood Association*. Stanford: Stanford University Press.

[9] Maccoby, E.E. (1990), Gender and relationships: a developmental account. *Amer. Psychol.* 45.4, 513–520, https://doi.org/10.1037//0003-066x.45.4.513

could be to superimpose habituation on imprinting. The net effect of superimposing one learning process on the other would be to produce a sharply peaked preference for something a bit different from the familiar. The evolutionary benefits and costs of inbreeding are likely to be numerous and to vary in importance from one species to the next. Therefore the precise balance will almost certainly differ between species. Even within a species, the balance is likely to depend on local conditions and on how inbred a population has become. Finally, the sexes may differ, especially when one sex is likely to have more matings than the other. Nonetheless, a general point remains. When choosing a mate an animal may have to pay careful attention, among other things, to similarities between its proposed partner and close kin.

Incest taboos

Many authors have suggested that individuals may derive reproductive success from incest taboos.[10] Those individuals who impose the prohibitions do not derive immediate personal benefits from them. Social benefits may be derived because the group does not have to pay the costs of caring for individuals who in various ways are less fit. An attempt to mount a purely eugenic argument would be confused because the maladaptive genes expressed when inbreeding is common are not removed from the population by preventing inbreeding. Indeed, inbreeding is the best way of getting rid of those genes in the long run.

Whether or not people are aware of the effects of inbreeding is another issue. In many cultures they are.[11] Awareness of the ill-effects of inbreeding would be best translated into the conviction that the aware individual should not have children with his or her sibling. Nothing more is required of Darwinian evolution. The awareness does not immediately translate into a conviction that others should be stopped from having children with their siblings.

---

10   Many authors have linked avoidance of inbreeding with the incest taboo. One of the most prominent is Wilson, E.O. (1998), *Consilience: The Unity of Knowledge*. New York: Alfred A. Knopf.
11   Durham, W.H. (1991), *Coevolution: Genes, Culture and Human Diversity*. Stanford: Stanford University Press.

Did inbreeding avoidance and incest taboos evolve by similar mechanisms or do they have a common utility in modern life? Incest taboos need not necessarily serve the same function as the inhibitions derived from early experience. What other mechanism for the cultural evolution of incest taboos should be entertained? Humans might often have an inclination to prevent other people behaving in ways in which they would not themselves behave. On this view, left-handers were in the past forced to adopt the habits of right-handers because the right-handers found them disturbing. Similarly the moral repugnance that many people show for homosexuality between consenting adults is often a violent one — in some societies homosexuality may be punished by death. In the same way, those who were known to have had sexual intercourse with close kin were discriminated against. People who had grown up with kin of the opposite sex were generally not attracted to those individuals, and disapproved when they discovered others who were. On this view, the incest taboo was nothing to do with society not wanting to look after the cognitively challenged offspring of inbreeding, since in many cases they had no idea that inbreeding was the cause. Rather, the disapproval was about suppressing abnormal behaviour, which is potentially disruptive in small societies. Such conformity looks harsh to modern eyes, even though plenty of examples of it are found in contemporary life.

When so much depended on unity of action in the environment in which humans evolved, wayward behaviour could have destructive consequences for everybody. It is not difficult to see why conformity should have become a powerful trait in human social behaviour.[12] Once in place, the desire for conformity, on the one hand, and the reluctance to inbreed, on the other, would have combined to generate social disapproval of inbreeding. The emergence of incest taboos would take on different forms, depending on which sorts of people, non-kin as well as kin, were likely to be familiar from early life.

In the Anglican *Book of Common Prayer* there is a table of Kindred and Affinity 'wherein whosoever are related are forbidden by the Church of England to marry together'. A man may not marry his mother, sister or daughter and a variety of other genetically related individuals. The

---

12   Westermarck, E. (1891), *The History of Human Marriage*. London: Macmillan.

restrictions for a woman are reciprocal and also, among others, she must not marry an uncle, nephew, grandparent or grandchild. The list continues with the following exclusions: a man may not marry his wife's father's mother or his daughter's son's wife. At least six of the 25 types of relationship that preclude marriage involve no genetic link. The Church of England did not worry about marriages between first cousins. Other cultures do, but here again striking inconsistencies are found. In a great many cultures marriages between first cousins who have parents of the same sex are forbidden whereas marriages between cousins who have parents of different sexes are not only allowed but, in many cases, actively encouraged.

If these ideas are correct, human incest taboos did not arise historically from a deliberate intention to avoid the biological costs of inbreeding. Rather, in the course of history, two separate mechanisms appeared. One was a developmental process concerned with striking an optimal balance between inbreeding and outbreeding when choosing a mate. The other was concerned with social conformity.[13] When these two propensities were put together, the result was social disapproval of those who chose partners from within their close family. When social disapproval was combined with language, verbal rules appeared which could be transmitted from generation to generation, first by word of mouth and later in written form.

Conclusions

While the incest taboo is not likely to be a result of Darwinian evolution, the preference for a mate slightly different from close kin undoubtedly is. Other factors also influence mate choice, particularly in humans. The attractiveness of a potential mate, the resources he or she might hold, and so forth also play a part. But in terms of producing grand-offspring, optimal outbreeding is important. The mating preference appears to be well designed to achieve that except in modern human society where having a lot of grandchildren is not usually important.

---

13   Westermarck's ideas are discussed at length in the chapters in Wolf, A.P. & Durham, W.H. (eds.) (2004), *Inbreeding, Incest and the Incest Taboo*. Stanford: Stanford University Press.

8. Genes in Development and Evolution

For the Darwinian evolutionary mechanism to work, something must be inherited with fidelity. Even if a single change in DNA provides the basis for a distinctive beneficial character of an individual, that is not sufficient for the development of the character. This gets to the heart of a lively debate in biology.[1] Genes have been defined in many different ways: as units of physiological function, units of recombination, units of mutation, or units of evolutionary process — when they have sometimes been imbued with 'selfish' intentions in order to help the understanding of the complexities of evolution. The problem of definition has been made worse as it has become clear that the same molecule of DNA may serve in processes that differ in function. In the post-genomic era, the emerging concepts of the gene pose a significant challenge to conventional assumptions about the relationship between

---

1   This debate is well-described in Noble, D. (2016), *Dance to the Tune of Life: Biological Relativity*. Cambridge: Cambridge University Press. Denis Noble argues that living organisms operate at multiple levels of complexity and must therefore be analysed from a multi-scale, relativistic perspective. Noble explains that all biological processes operate by means of molecular, cellular and organismal networks. The interactive nature of these fundamental processes is at the core of biological relativity and, as such, challenges simplified molecular reductionism. Noble shows that such an integrative view emerges as the necessary consequence of the rigorous application of mathematics to biology. Drawing on his pioneering work in the mathematical physics of biology, he shows that what emerges is a deeply humane picture of the role of the organism in constraining its chemistry, including its genes, to serve the organism as a whole, especially in the interaction with its social environment. This humanistic, holistic approach challenges the common gene-centred view held by many in modern biology.

genome structure and function, and between genotype and developed characteristics.[2]

The word 'gene' does not, then, have a clear unambiguous meaning.[3] For some scientists it meant simply a sequence of DNA, for others it referred specifically to those segments of DNA that are transcribed into ribonucleic acid (RNA) and then translated into a protein. By contrast, many segments of RNA — the so-called non-coding RNAs — have regulatory functions, and the term 'gene' is extended by many molecular geneticists to include the DNA sequences coding for these RNAs. These different meanings of gene sometimes get conflated, with subsequent confusion of thought.

Despise the semantic confusion, the use of variations in DNA to identify individuals is crucially important in the forensic analysis of crimes. In other areas of biology the variation is used to establish relationships between species and the probable evolution of taxonomic groups. Moreover the technologies for producing new types of crops that are resistant to disease or water shortage are now well developed. Scientists collaborating on the Human Genome Project have elucidated nearly all the DNA sequences on all 23 pairs of chromosomes found in a human cell.[4] It is a staggering achievement. Much epidemiological research in recent years has been based on sequencing the entire human genome and looking at mutant alleles that correlate with disease. A surprising result of these genome-wide association

The process of making a cake. Photo by Roozitaa (2012), Wikimedia, https://commons.wikimedia.org/wiki/File:Making_Chiffon_cake_2.jpg, CC-BY 3.0.

---

2   See Sultan, S. (2015), *Organism and Environment: Ecological Development, Niche Construction, and Adaptation.* Oxford: Oxford University Press, http://dx.doi.org/10.1093/acprof:oso/9780199587070.001.0001

3   Keller, E.F. (2000), *Century of the Gene.* Cambridge, MA: Harvard University Press. An accessible and clear-headed introduction to genetics is given in Griffiths, P. & Stotz, K. (2013), *Genetcs and Philosophy: An Introduction.* Cambridge: Cambridge University Press.

4   Many of the DNA strands come in many different forms that provide the basis for forensic studies and attempts to discover genetic relationships between people.

studies has been that, even when large populations are studied, and the disease of interest is common, such as Type 1 diabetes, few significant genetic effects are found and the effects of any one specific difference in DNA are generally small. Single-gene effects are unusual and largely restricted to relatively rare diseases, such as phenylketonuria or haemophilia. The excitement about what is being done should be greatly moderated. 'The Book of Life', as one leading scientist called it, does not provide the complete story about human nature.

Genes in development

The starting points of development include the genome which provides information of a kind.[5] They also include factors external to the genome; the social and ecological conditions in which the individual grows up are crucial. A low-tech cooking metaphor serves to shift the focus onto the multi-causal and conditional nature of development. Using butter instead of margarine may make a cake taste differently when all the other ingredients and cooking methods remain unchanged. But if other combinations of ingredients or other cooking methods are used, the distinctive difference between a cake made with butter and a cake made with margarine may vanish. Similarly, a baked cake cannot readily be disaggregated into its original raw ingredients and the various cooking processes, any more than a behaviour pattern or a psychological characteristic can be disaggregated into its genetic and environmental influences and the developmental processes that gave rise to it. In the cooking analogy, the raw ingredients represent the many genetic and environmental influences, while cooking represents the biological and psychological processes of development. Nobody expects to find all the separate ingredients represented as discrete, identifiable components in a cake. Similarly, nobody should expect to find a simple correspondence between a particular gene (or a particular experience) and particular aspects of an individual's behaviour or personality.

---

5   The use of the term information exclusively applied to genes has been sharply criticised by Oyams, S. & Lewontin, R. (2000), *The Ontogeny of Information*. 2nd ed. Durham, NC: Duke University Press, https://doi.org/10.1215/9780822380665. As the title of her book suggests, all factors impinging on the developing organism provide information of a kind.

The language of a gene 'for' a particular characteristic is exceedingly muddling to the non-scientist — and, if the truth be told, to some scientists as well. What the scientists mean (or should mean) is that a genetic difference between two groups is associated with a difference in a characteristic. They know perfectly well that other things are important and that, even in constant environmental conditions, the outcome depends on a combination of many genes. Particular combinations of genes have particular effects, and a gene that fits into one combination may not fit into another. Unfortunately, the language of a gene 'for' a characteristic has a way of sometimes seducing scientists themselves into believing their own sound-bites. Such language rests on a profound misunderstanding.

The notion that genes are simply blueprints for an individual human is hopelessly misleading. In a blueprint, the mapping works both ways: starting from a finished house, a room can be found on the blueprint, just as the room's position is determined by the blueprint during the building process. This straightforward mapping is not true for genes and individual human behaviour patterns, in either direction.

The common image of a genetic blueprint for behaviour fails because it is too static, suggesting that adult organisms are merely expanded versions of the fertilized egg. In reality, developing organisms are dynamic systems that play an active role in their own development. Even when a particular base in a strand of DNA or a particular experience is known to have a powerful negative effect on the development of behaviour, biology has an uncanny way of finding alternative routes. If the normal developmental pathway to a particular form of adult behaviour is impassable, another way may often be found. The individual may be able, through its behaviour, to match its environment to suit its own characteristics.

Strands of DNA do not, on their own, make behaviour patterns or physical attributes. They code for polypeptides, the precursors of proteins or small molecules of RNA. The proteins are crucial collectively to the functioning of each cell in the body. Some proteins are enzymes, controlling biochemical reactions, while others form the physical structures of the cell. These protein products of genes do not work in isolation, but in a cellular environment created by local conditions.

The DNA content of an individual organism can be measured accurately. When the amount of DNA is compared to the relative size of the nervous system and the complexity of behaviour it generates, a lack of correlation is surprising. Mice have 6000 times as much DNA as bacteria which makes sense if the differences in the DNA are responsible for the differences in complexity. Humans have no more DNA than mice. In other words, whatever else happened in evolution, the gradual emergence of behavioural complexity within the mammals was not achieved by accumulating the genes that code for protein components. Some of the lack of association may be explained by the number of genes not being the same as the amount of DNA. It may also be attributed to 'junk' DNA which, it used to be supposed, had no effect on the organism's developed characteristics. The discovery that some or all genes in this so-called junk code for small molecules of RNA has profound implications for the regulation of development. At the very least, it means that the growth of nervous systems and the emergence of behaviour are critically dependent on regulation and the combinatorial action of genes. This means that the correspondence between genes and behaviour is never likely to be simple. Many genes really do code for polypeptides, of course, but they represent nothing else since the process of development is generative. A small subset of genes and cytoplasmic conditions start the whole process after fertilisation of the egg. These starting conditions create products that switch off some active genes, switch on others and bring the developing components into contact with new influences from outside. And so the whole process continues until death.

It is clear, then, that because of the system in which they are embedded, no simple correspondence is found between individual genes and particular behaviour patterns or psychological characteristics. Genes do not code for parts of the nervous system and they certainly do not code for particular behaviour patterns. Any one aspect of behaviour is influenced by many genes, each of which may have a big or a small effect. Conversely, changes in any one of many genes can have a major disruptive effect on a particular aspect of behaviour. A disconnected wire can cause a car to break down, but this does not mean that the wire by itself is responsible for making the car move.

Without a strong set of binding ideas, it isn't easy to think about all aspects of the various strands of evidence, which often seem to point in opposite directions. Some theorists have argued that the seemingly simple and orderly characteristics of development (such as they are) are generated by dynamic processes of great complexity. Many mathematical techniques, such as catastrophe theory and 'chaos', have been developed to deal analytically with the complexities of dynamical systems. A promising empirical approach is collecting evidence across different levels of analysis.

Heritability

When offspring look like their parents or other members of their family, it is reasonable to assume that they have inherited something. The total set of characteristics that are inherited are rarely correlated. So in human families a boy might have the big nose of his uncle, the ginger hair of his father and the retiring disposition of his grandmother. Mendelian inheritance and the recombination of characteristics in each generation explains why this should be so. Also, many different processes might be involved; most are genetic but some are environmental. Attempts to sort out the different possibilities have risen to the concept of heritability. Instead of asking whether a child's characteristics are caused by genes or caused by the environment, the question instead became: 'How much is due to each?' In a more refined form, the question is posed thus: 'How much of the variation between individuals in a given character is due to differences in their genes, and how much is due to differences in their environments?'

The meaning of heritability is best illustrated with an uncontroversial characteristic such as height, which is clearly influenced by both the individual's family background (genetic influences) and nutrition (environmental influences). The variation between individuals in height attributable to variation in their genes may be expressed as a proportion of the total variation within the population sampled. This index is known as the heritability ratio. The higher the figure, which can vary between 0 and 1.0, the greater the contribution of genetic variation to individual variation in that characteristic. So, if people differed in height solely because they differed in their genes, the heritability of height would be 1.0; if, on the other hand, variation in height arose entirely from individual differences in environmental factors such as nutrition then the heritability would be zero.

Calculating a single number to describe the relative contribution of genes and environment has obvious attractions. Estimates of heritability are of undoubted value to animal breeders, for example. Given a standard set of environmental conditions, the genetic strain to which a pig belongs will predict its adult body size better than other variables such as the number of piglets in a sow's litter. If the animal in question is a cow and the breeder is interested in maximising its milk yield, then knowing that milk yield is highly heritable in a particular strain of cows reared under standard rearing conditions is important.

Behind the deceptively plausible ratios lurk some fundamental problems. For a start, the heritability of any given characteristic is not a fixed and absolute quantity — tempted though many scientists have believed otherwise. Its value depends on a number of variable factors, such as the particular population of individuals that has been sampled. For instance, if heights are measured only among people from affluent backgrounds, then the total variation in height will be much smaller than if the sample also includes people who are small because they have been undernourished. The heritability of height will consequently be larger in a population of exclusively well-nourished people than it would be among people drawn from a wider range of environments. Conversely, if the heritability of height is based on a population with relatively similar genotypes — say, native Icelanders — then the figure will be lower than if the population is genetically more heterogeneous; for example, if it includes both Icelanders and African Pygmies. Thus, attempts to measure the relative contributions of genes and environment to a particular characteristic are highly dependent on who has been measured and in what conditions.

Another problem with heritability estimates is that they reveal nothing about the ways in which genes and environment contribute to the biological and psychological processes of development. This point becomes obvious when considering the heritability of a characteristic such as 'walking on two legs'. Humans walk on fewer than two legs only as a result of environmental influences such as war wounds, car accidents, disease or exposure to toxins before birth. In other words, all the variation within the human population results from environmental influences, and consequently the heritability of 'walking on two legs' is zero. And yet walking on two legs is clearly a fundamental property of being human, and is one of the more obvious biological differences between humans and other great apes such as chimpanzees or gorillas. It obviously depends heavily on genes, despite having a heritability of zero in humans. A low heritability clearly does not mean that development is unaffected by genes.

The effects of a particular set of genes depend critically on the environment in which they are expressed, while the effects of a particular sort of environment depend on the individual's genes. Even in animal breeding programmes that use heritability estimates to practical advantage, care is still needed. If breeders wish to export a particular genetic strain of cows that yields a lot of milk, they would be wise to check that the strain will continue to give high milk yields under the different environmental conditions of another country.

Epigenetics

The often uncanny similarities between identical twins provide striking evidence for the importance of genes in shaping physical and behavioural characteristics. On the other hand identical twins can differ markedly from each other.[6] The cues that come from the environment are often those that regulate the regulators. Much of the plasticity seen in development is generated this way. The course of an individual's development may be radically different depending on the nature of these cues. Individuals with identical genomes do not necessarily have identical adult characteristics. In the case of schizophrenia for instance one identical twin may develop the disease while the other does not.

Identical twins reared apart are sometimes more like each other than those reared together.[7] To put it another way, rearing two genetically identical individuals in the same environment can make them less similar. This fact pleases neither the extreme environmental determinist nor the extreme genetic determinist. The environmental determinist supposes that twins reared apart must have different experiences and should therefore be more dissimilar in their behaviour than twins who

---

6 See Spector, T. (2012), *Identically Different: Why You Can Change Your Genes*. London: Weidenfeld & Nicolson.

7 Shields, J. (1962), *Monozygotic Twins Brought up Apart and Brought up Together*. Oxford: Oxford University Press.

grow up together in the same environment. The genetic determinist does not expect to find any behavioural differences between genetically identical twins reared together; if they have had the same genes and the same environment, then how can they be different? Of course, one twin provides a social environment for the other and often one sibling will not do what the other one is doing.

All processes involved in development have been subsumed under the heading of epigenetics. In a restricted sense, epigenetic processes are those that result in the silencing or activation of gene expression through such modification of the roles of DNA or its associated RNA and protein. The term has therefore come to describe, for many, those molecular mechanisms through which both dynamic and stable changes in gene expression are achieved, and ultimately how variations in extracellular input and experience by the whole organism of its environment can modify regulation of DNA expression. Some authors continue to use this broader definition of epigenetics to describe all the developmental processes, behavioural and physiological as well as molecular, that bear on the character of the organism.[8] In all these

---

[8] The developmental processes involved were subsumed under the general heading of 'epigenetics' by Waddington, C.H. (1957) in *The Strategy of the Genes* (London: Allen & Unwin). He distinguished this term from the eighteenth-century term 'epigenesis', which had been used to oppose the notion that all the characteristics of the adult were preformed in the embryo. More recently, epigenetics has become mechanistically defined as the molecular processes by which traits defined by a given profile of gene expression can persist across mitotic cell divisions, but which do not involve changes in the nucleotide sequence of the DNA (see Carey, N. (2012). *The Epigenetics Revolution: How Modern Biology is Rewriting Our Understanding of Genetics, Disease and Inheritance*. London: Icon Books Ltd.). The general principles apply at higher levels of organisation and are involved in mediating many aspects of developmental plasticity seen in intact organisms. For that reason, some authors continue to use Waddington's broader definition of epigenetics to describe all the developmental processes that bear on the character of the organism. The processes involved in gene expression and suppression can be transmitted from one generation to the next (See Gissis, S.B. & Jablonka, E. (2011), *Transformations of Lamarckism: From Subtle Fluids to Molecular Biology*. Cambridge, MA: MIT Press, and Miska, E.A. & Ferguson-Smith, A.C. (2016), Transgenerational inheritance: models and mechanisms of non-DNA sequence-based inheritance, *Science* 354. 6308, 59–63, https://doi.org/10.1126/science.aaf4945). Further support for the revision of the orthodoxies of evolutionary theory has come from microbiology (Shapiro, J. (2011), *Evolution: A View from the 21st Century*. Upper Saddle River, NJ: FT Press Science). Shapiro argues that cells must be viewed as complex systems that control their own growth, reproduction and shape their own evolution over time. He referred to it as a 'systems engineering' perspective and noted interestingly that, 'Most of the interactions between biomolecules tend to be relatively weak and need multiple synergistic attachments to produce stable functional complexes' (Shapiro, 2011: 31).

usages, epigenetics usually refers to what happens within an individual developing organism. Whether a broad or restricted view of epigenetics is taken, the discovery of epigenetic phenomena has led to a revolution in thinking about the importance of developmental processes.

The molecular processes involved in the development of an organisms characteristics were initially worked out for the regulation of cellular and proliferation. All cells within the body contain the same genetic sequence information, yet each lineage has undergone specialisation to become a skin cell, hair cell, heart cell, and so forth. These differences within a developing individual are inherited from mother cells to daughter cells. The process of differentiation involves the expression of particular genes for each cell type in response to cues from neighbouring cells and the extracellular environment, with the silencing of others. Genes that have been silenced at an earlier stage remain silent after each cell division (except in cancers). Such gene silencing provides each cell lineage with its characteristic pattern of gene expression. These epigenetic marks are faithfully duplicated across each cell division, stable cell differentiation results and serves many different functions.

Molecular mechanisms are involved in the activation or silencing of genes. One of the silencing mechanisms involves a process known as methylation. Chromosomes consist of strands of chromatin. DNA is organized along chromatin in packets known as nucleosomes. These have a molecule with a hydrogen atom on one of its arms. If this is replaced by a methyl group, the nucleosomes close up and the DNA is less able to be expressed as messenger RNA which in turn forms the template for synthesising protein. Conversely if the methyl group is replaced by a hydrogen atom, the DNA on the affected nucleosomes can be expressed.

An important mechanism in development involves small molecules of non-coding micro-RNA. When these small molecules are expressed they may bind onto messenger RNA which links as an intermediate between DNA and protein, with the result that the gene that expressed the messenger RNA loses its capacity to code for protein and is effectively silenced. The regulators have themselves to be regulated, and unraveling the networks will take a great deal of research, but the general principles involved in producing differences in cell lines are already apparent.

Many examples in biology demonstrate the dependence of gene expression on local conditions. After a fire on the high grassland planes of East Africa, for example, the young grasshoppers are black instead of being the normal pale yellowish-green. Something has switched the course of their development onto a different track. The grasshopper's colour makes a big difference to the risk that it will be spotted and eaten by a bird, and the scorched grassland may remain black for many months after a fire. So matching its body colour to the blackened background is important for its survival. The developmental mechanism for making this switch in body colour is automatic and depends on the amount of light reflected from the ground.[9] If the young grasshoppers are placed on black paper they become black when they moult to the next stage. But if they are placed on pale paper the moulting grasshoppers are the normal green colour. The grasshoppers actively select habitats with the colour that match their own. If the colour of the background changes they can also change their colour at the next moult to match the background, but once they reach adulthood they are committed to one colour. This striking example illustrates at the level of the whole organism the role of epigenetics in development.

Selfish genes

Turning to evolutionary processes, a crucial question is at what level of organisation does the process of Darwinian evolution act? The selfish gene approach made famous by Richard Dawkins[10] has been valuable in helping to understand self-sacrifice, and conflicts between the sexes and generations. The language of genes having metaphorical intentions helps people to deal with the complicated dynamics of evolution. Such explanations are not meant to be treated in the way usually employed by an experimental scientist; they provide a framework in which to think about phenomena that would otherwise be neglected.

---

9 Rowell, C.H.F. (1971), The variable coloration of the acridoid grasshoppers. *Adv. Insect Physiol.* 8, 145–198, https://doi.org/10.1016/s0065-2806(08)60197-6
10 Dawkins, R. (1976), *The Selfish Gene*. Oxford: Oxford University Press.

Most people get their minds around complex processes when they attribute intentions to them. It is a powerful way of thinking about systems. Weather forecasters, having to cope with explaining appallingly complex problems, make statements like: 'The depression is trying to move in from the west'. The language encourages thought about endpoints rather than with all the details of how they are achieved.

Is the gene the target of selection in evolution? It may be helpful to forget biology for a moment and think about the spread of a new brand of biscuit in supermarkets. Consider the spread from the perspective of the recipe. While shoppers select biscuits and eat them, it is the recipe for making desirable biscuits that survives and spreads in the long run. A phrase in the recipe might specify the amount of sugar to be added and makes the difference between a popular and a less popular biscuit. In that sense it is selfish. This novel way of looking at things is unlikely to mislead anyone into believing that what shoppers really do in supermarkets, when they pick a particular brand of biscuit off the shelves, is select a word in the recipe used for making the biscuits. They select the brand of biscuit they like.

Richard Dawkins author of *The Selfish Gene*. Photo by Marty Stone (2009), Wikimedia, https://commons.wikimedia.org/wiki/File:Richard_Dawkins_35th_American_Atheists_Convention.jpg, CC-BY 2.0.

Darwin used his metaphor of 'natural selection' because he was impressed by the ways in which plant and animal breeders artificially selected the characters they sought to perpetuate. The agents of differential survival and differential reproductive success will usually be characteristics of whole individuals including the structures they make, but they might be characteristics of molecules or of symbiotic groups, or the evolvability of taxonomic lineages.

The power of the selfish gene language has been used misleadingly to prop up the idea of the gene as 'programmer'. The mechanistic language does not translate into the teleological language. For population geneticists, a genetic difference is identified by means of a biochemical, physiological, structural or behavioural difference between organisms (after other potential sources of difference have been excluded by appropriate procedures). The popular language of genes' intentions and the more orthodox language of genetic differences are not simply alternative ways of describing the same thing. In the technically precise language of population geneticists, a genetic allele must be compared with another from which it differs in its consequences. In selfish-gene language, it stands alone as an entity, absolute in its own right. The perception generated by one meaning of gene does not relate to the same evidence as that generated by the other.

Conclusions

An important point, often made but equally often ignored, is that correlations between the behaviour patterns of the parent and those of the do not necessarily arise because they have genes in common. They may arise because other conditions that are necessary for the peculiarities of their behaviour are shared. Common odours and preferences for familiar smells might arise from the particular combination of bacteria that breakdown the fats secreted onto the body surface. When the bacteria pass from mother to offspring, so does the source of her special smell. This is not to downplay the roles of genes. But it emphasises that the nothing-but approach to genes is clearly wrong. Taking a systems approach to the role of genes generates much more fruitful understanding than treating them as providing single causes for development and evolution.

9. Active Role of Behaviour

The ability of animals to respond differentially to one of several options is an important part of adaptive behaviour. In colloquial terms they make a choice. Charles Darwin suggested that members of one sex choose to mate with individuals with a striking feature such as the tail of the male peacock. Choice can take many different forms. One is involved in predators' choice of prey. When gazelle see a predator like a cheetah they jump into the air, a behaviour pattern called 'stotting'. Cheetah seem to learn not to chase jumping gazelle.[1] A similar case is the small falcon, the merlin, which takes other small birds on the wing.[2] When it hunts flying skylarks, its potential prey start singing. The more the skylark sings, the more likely is the merlin to abandon the chase and hunt for other skylarks that don't sing so much. The merlin has probably learned that the skylark singing a lot is more vigorous and more difficult to catch and chooses to attack other individuals.

Apparent design emerges, even when it is at the end of the long and complicated process of development. Development depends on the constancy of many genetic and environmental conditions. If any of these conditions changes, as can happen to environmental conditions when organisms move away from the natal area, the characteristics of the organism can also change. High mobility by organisms would have

---

1   Fitzgibbon, C.D. & Fanshawe, J.H. (1988), Stotting in Thomson's gazelles: an honest signal of condition. *Behav. Ecol. Sociobiol.* 23.2, 69–74, https://doi.org/10.1007/bf00299889

2   Cresswell, W. (1994), Song as a pursuit-deterrent signal, and its occurrence relative to other anti-predation behaviours of skylark (*Alauda arvensis*) on attack by merlins (*Falco columbarius*). *Behav. Ecol. Sociobiol.* 34.3, 217–223, https://doi.org/10.1007/bf00167747

frequently placed them in conditions that revealed heritable variation not previously apparent in the population. By their mobility, in the case of animals, or facility to disperse their seeds in the case of plants, organisms would have exposed themselves to new conditions that might reveal heritable variability.

The environment does not simply set a problem to which the organism has to find a solution. The organism can do a great deal to select or create an environment to which it is best suited. For example beavers dam rivers, flood valleys and create private lakes for themselves. The concepts derived from such examples have been developed extensively and they are now referred to collectively as 'niche construction'.[3] The effects of behavioural control can be especially great when a major component of the environmental conditions with which animals have to cope is provided by their social environment. When individuals compete with each other within a social group, the outcome of the competition depends in part on each individual's capacity to predict what the other will do.

Lake created by beavers damming a river (2005), Wikimedia, https://commons.wikimedia.org/wiki/Algonquin_Provincial_Park#/media/File:Biberdamm_2_db.jpg, Public Domain.

---

3   Laland, K.N., Odling Smee, J. & Gilbert, S.F. (2008),   Evo-devo and niche construction: building bridges. *Journal of Experimental Evolution* 310B, 549-566.

Play behaviour, prominent in young mammals and some birds, is spontaneous and rewarding to the individual; it is intrinsically motivated and its performance serves as a goal in itself. The player is to some extent protected from the normal consequences of serious behaviour. The behaviour appears to have no immediate practical goal or benefit. Playing with other individuals may be preceded or accompanied by specific signals or facial expressions indicating that the behaviour is not to be taken as a threat. Play is the antithesis of 'work' or 'serious' behaviour. The behaviour consists of actions and, in the case of humans, thoughts, expressed in novel combinations. When playing with others, a normally dominant individual may become temporarily subordinate. Individual actions or thoughts are performed repeatedly; they may also be incomplete or exaggerated relative to non-playful behaviour in adults; play looks different. Play is sensitive to prevailing conditions and occurs only when the player is free from illness or stress. Play is an indicator of well-being. Playful play is accompanied by a particular positive mood state in which the individual is more inclined to behave in a spontaneous and flexible way.[4]

When young animals playfully practise the stereotyped movements they will use in earnest later in life, they are likely to improve the coordination and effectiveness of these behaviour patterns. The short dashes and jumps of a young gazelle when it is playing bring benefits that may be almost immediate, as it faces the threat of predation from

---

4   A full discussion of play is given in Bateson, P. & Martin, P. (2013), *Play, Playfulness, Creativity and Innovation*. Cambridge: Cambridge University Press. 'Playfulness' is a positive mood state that facilitates and accompanies 'playful play', a subset of broadly defined play. A distinction is drawn between playful play and non-playful play. Playfulness, the defining feature of playful play, is a positive mood state that is not always detectable in observable behaviour. The behaviour of a playful human is captured by numerous synonyms, including cheerful, frisky, frolicsome, good-natured, joyous, merry, rollicking, spirited, sprightly and vivacious. Some of these terms relate to human emotions that could not be readily identified in animals without much anthropomorphic projection. Some, though, are descriptive of visible behaviour and can be defined ostensively, such as when two kittens engage vigorously in social play. In animals, as in humans, playfulness may be inferred from the context in which it occurs. What the animals do may vary — from playing with objects at one moment to playing with another individual at the next — but the playful state underlying their behaviour is the same.

cheetah or other carnivores and needs considerable skill when escaping.[5] Even though the benefits may be immediate in such cases, they may also persist into adult life.

Many theories of the functions of play have continued to focus on its role in enabling the developing individual to acquire and practise complex physical skills and, by so doing, fine tune neuromuscular systems. Others theories, derived from observing how much young animals play with each other, have emphasised how individuals also develop social skills and cement their social relationships; play may also serve to improve the individual's capacity to compete and cooperate with other members of their own species. Play can make an individual more resistant to stress, and enlarge its repertoire. Play may enhance an individual's resourcefulness and flexibility and make it able to adjust to new conditions. Play may enhance its ability to cooperate with others and to co-exist with older members of its own species. Play may increase its knowledge of its home range. Play, or at least some components of it, allows young animals to simulate, in a relatively safe context, potentially dangerous situations that will arise in their adult life. They learn from their mistakes, but do so in relative safety. On this view, play exerts its most important developmental effects on risky adult behaviour such as fighting, mating in the face of serious competition, catching dangerous prey, and avoiding becoming someone else's prey. Indeed, the behaviour patterns of fighting and prey-catching are especially obvious in the play of cats and other predators, whereas safe activities such as grooming, defecating and urinating have no playful counterparts.

Dog soliciting play with a bow. Photo by Thomas Zimmermann (2015), Wikimedia, https://commons.wikimedia.org/wiki/File:Vorderkoerpertiefstellung_THWZ.jpg, CC-BY 3.0.

---

5 Gomendio, M. (1988), The development of different types of play in gazelles: implications for the nature and functions of play. *Anim. Behav.* 36.3, 825–836, http://dx.doi.org/10.1016/S0003-3472(88)80165-9

When differences between the sexes arise in play, as they often do, these are reflected in differences between the sexes in the activities of adults. For instance young female chimpanzees seem to behave maternally towards sticks, doing so much more than males and ceasing to do so when they have real offspring to care for.[6] One study showed that stick-carrying consisted of holding or cradling detached sticks pieces of bark, small logs or woody vine with their hand or mouth, underarm or, most commonly, tucked between the abdomen and thigh. Individuals sometimes carried sticks for periods of up to four hours or more during which they rested, walked, climbed, slept and fed as usual. The occurrence of stick-carrying peaked among juveniles and was higher in females than males. This sex difference could not be explained by a general propensity for females to play with objects more than males, because several types of object such as weapons were played with more by males. Males in many species, including humans, perform more rough and tumble play than females and engage in more violent activities when adult.

Siberian tiger mum and daughter play-fighting. Photo by Tambako The Jaguar (2011), Flickr, CC BY-ND 2.0, https://www.flickr.com/photos/tambako/6831507351

Play has features that are likely to make it suitable for finding the best way forward in a world of conflicting demands. In acquiring cognitive skills, individuals are in danger of finding sub-optimal solutions to the many problems that confront them. In deliberately moving away from what might look like the final end state, each individual may arrive somewhere that is better. Play may therefore fulfil an important probing role that enables the individual to escape from false end-points or 'local optima'. An analogy is a mountain surrounded by lesser peaks.

---

6 Kahlenberg, S.M. & Wrangham, R.W. (2010), Sex differences in chimpanzees' use of sticks as play objects resemble those of children. *Current Biol.* 20.24, R1067-R1068, https://doi.org/10.1016/j.cub.2010.11.024

A climber might get to the top of a lesser peak only to discover that he or she had to descend before scaling a higher one. When the metaphorical climber is on a lower peak, active ways of getting off it can be highly beneficial. In practice what this could mean that the activities involved in play discover possibilities that are better than those obtained without play.

All short-term quantitative studies of play in animals find that some individuals play more than others. In humans, five main dimensions have been used to describe the variation in personalities. Many of these are not usually regarded as attributes of cognitive ability, but the dimension 'Openness to Experience' is one that could have developed as the result of play. The descriptions of people on one dimension range from Creativity to Analytical Ability. In a survey of humans, the individuals who believed that they were playful also believed that they were creative.[7] Respondents were asked to offer ideas for the uses of two items, a jam jar and a paperclip. In the literature on creativity, those individuals who produce few answers are referred to as 'convergers' and those who produce many suggestions are known as 'divergers'. The typical sole response from a converger when asked for uses for a paper clip was 'Clip paper together'. The response from one diverger in the survey was: 'Clip papers, unfold to clean fingernails, general clothes fixing in an emergency, put on a magnet for a science experiment for children, make a mobile with lots of them, make a sculpture with one or more of them, earrings, pick a lock'. If there had no been a cut-off after ten answers, this person would probably have gone on. Most of the respondents provided a relatively small number of uses for the objects and only a few offered many uses. The respondents who regarded themselves as playful and producers of new ideas were much more likely to give lots of uses.

The differences between individuals might reflect the variation in almost every character of body and behaviour. It might instead (or in addition) reflect the benefits of being different from others. In a population that consists mostly of females it is advantageous to be a male — and vice versa. In cooperative species, providing a particular set of skills may complement a different set possessed by others. In humans

---

7   Bateson, P. & Nettle, D. (2014), Playfulness, ideas, and creativity: a survey. *Creativity Research J.* 26.2, 219–222, https://doi.org/10.1080/10400419.2014.901091

those who suggest new ways of looking at the world are complemented by others who put such ideas to good use. Creativity and innovation are mutually beneficial but not necessarily found in the same person.[8]

Environmental change

The environment does not cease to be important even if it normally remains constant. Change the environment and the outcome of an individual's development may be utterly different. Indeed, if an individual does not inherit its parents' environment along with their genes and other transmittable factors, it may not be well adapted to the conditions in which it now finds itself. Its behaviour may enable it to cope.

A rule for learning, or for any other kind of developmental process, is not simply a gene written large. A straightforward correspondence between a gene and a rule for changing behaviour is no more likely than a straightforward relationship between gene and behaviour (see Chapter 8). The same point applies with equal force to all the other epigenetic rules that bring order to development. Presumably, if the rules have any universality in natural conditions, the experience that affects them must be a common feature of all the animals having those rules. Alternatively, they must be well buffered against change by the particular conditions in which an individual finds itself.

When behaviour changes in response to alterations in the environment, it seems likely that the specific ways in which animals tune their behaviour to local conditions are themselves the products of Darwinian evolution. If rules for learning fit the animal's information-gathering equipment to particular problems and which may have been subject to Darwinian evolution, then the conditions necessary for their development must pass in some way from one generation to the next. While the mode of transmission may commonly involve genes, the rules for modifying behaviour are hardly likely to spring fully armed out of the genome. Criticism of the assertion that genes code for behaviour that is not learned applies just as forcibly to the rules that are involved

---

8   The distinction between creativity and innovation is emphasised in Bateson, P. & Martin, P. (2013), *Play, Playfulness, Creativity and Innovation*.

in the development of behaviour. Such rules represent the workings of an already functional nervous system and body. They themselves have to develop and depend on structures that require for their development a complex interplay between the products of many genes and many conditions external to the genome.

Many features of the inferred rules for learning seem to be profoundly modified by experience. For instance, whether or not initially neutral cues are treated as potentially relevant or ignored is greatly affected by the animal's prior history. Such selectivity in responsiveness to external conditions can be of great use to the animal. In many experiments on associative learning monkeys are rewarded with food. The machines that dispense the food are commonly designed to drop peanuts into a cup when the monkey is to be rewarded. Many monkeys do not like the peanuts at first. They have to be deprived of their regular food and accustomed to the peanuts for weeks before they will take them with any readiness, let alone treat the nuts as rewards for appropriate behaviour. In such cases, which are not exceptional, experience expands the range of what the monkeys regard as acceptable food.

It might be argued that a spontaneously expressed rule could still be detected at work behind the scenes, since the general category of food, and its effectiveness as a reward, was in some sense built in. In other cases, though, it becomes more difficult to pinpoint what might or might not act as a reward without extensive knowledge of the animal's previous experience. For instance, the condition in which it becomes possible for an animal to perform an act that would bring it food becomes rewarding in itself. So the animal will work in order to provide itself with those conditions. In this way, lengthy chains of behaviour can be developed with one event providing the terminating condition for one action and the enabling condition for the next. This is the basis for many complex circus acts performed by animals.

An adaptation of an animal's behaviour to the environment in which it lives gives the appearance of good design. The adaptations are often the result of Darwinian evolution, but the adaptability of organisms will mean that the adaptations may have developed during the lifetime of the individual. An individual whose body has been damaged in an accident or who is burdened with a mutation that renders its body radically different from other individuals may be able to accommodate to such abnormality. In doing so, the individual may develop novel structures and behaviour not seen in other individuals of the same species. Such accommodation can be particularly marked when it occurs early in development. A goat born without forelimbs walked about on its hind legs and developed a peculiar musculature and skeleton. A modern instance is a bipedal domestic dog.[9] The animals have coped with an abnormality by accommodating to it, producing coordinated changes in functionally related characters. Similarly, humans born with limb abnormalities as a result of exposure to a teratogen such as thalidomide develop strategies to cope, for example, by handling objects using their feet or teeth in ways for which others might use their hands.

The capacity of the individual to respond to neural damage is remarkable, particularly when the damage occurs early in life. In such cases described in humans, the brain reorganises and morphologically can look markedly different from the brain of a normal individual. Even so, the effects on behaviour may be scarcely detectable and the plasticity at the neural level may be accompanied by robust development at the behavioural level.

Another form of 'coping', found especially during early development, arises when the organism must make immediate responses to survive a challenge but, in contrast to accommodation responses, the normal developmental sequence is not necessarily disrupted. Although these responses may involve either structural or temporal changes in the course of development, they do not entail a fundamental change in the normal pattern of development. Thus, the phenotypic consequences are not as marked as those that involve accommodation, but they may have a costs and become disadvantageous to the individual later in life.

Conclusions

Individuals differ for a variety of reasons, some genetic and some stochastic. Undoubtedly their plasticity, which comes in many forms, also contributes greatly to the variation commonly found in most populations. The processes involved in plasticity can operate at many

---

9   The remarkable ability of Faith, the two-legged dog, can be viewed on YouTube https://www.youtube.com/watch?v=5QKG3CKZTYU

different levels, ranging from the molecular to the behavioural, some involving adaptability to what may be novel challenges and some responding conditionally to local circumstances. Differences between individuals can be triggered in a variety of ways, some mediated through the parent's characteristics. Sometimes phenotypic variation arises because the environment triggers a developmental response that is appropriate to those ecological conditions. Sometimes the organism 'makes the best of a bad job' in suboptimal conditions. Sometimes the buffering processes of development may not cope with what has been thrown at the organism, and a bizarre phenotype is generated. Whatever the appearance of a well-designed organism's characteristics, the various forms of plasticity illustrate why individuals of the same species can differ so much.

10. Adaptability in Evolution

Compelling examples of the interplay between genes and the environment may be found throughout the animal and plant kingdoms (see Chapter 8). Genetically identical individuals may develop in different ways, depending on environmental cues they received when they were young.

Well-meaning attempts to break out of the nature-nurture straitjacket have often resulted in a bewildering portrayal of development as a process of impenetrable complexity. Indeed, development seemed so unfathomably complex to eighteenth-century biologists that they believed that it must depend on supernatural guidance. On the surface the processes involved in behavioural development do indeed look forbiddingly complicated.

Order underlies even those learning processes that make individuals different from each other. Knowing something of the underlying regularities in development does bring an understanding of what happens to the child as it grows up. The ways in which learning is structured, for instance, affect how the child makes use of environmental contingencies and how the child classifies perceptual experience. Yet predicting precisely how an individual child will develop in the future from knowledge of the developmental rules for learning is no easier than predicting the course of a chess game. The rules influence the course of a life, but they do not determine it. Like chess players, children are active agents. They influence their environment and are in turn affected by what they have done. Furthermore, children's responses to new conditions will, like chess players' responses, be refined or embellished as they gather experience. Sometimes normal development of a particular ability requires input from the environment at a particular time; what happens next depends on the character of that input. The upshot is that, despite their underlying regularities, developmental processes seldom proceed in straight lines. Big changes in the environment may have no effect whatsoever, whereas some small changes may have big effects.

A more general point is that the development of individuals is readily perceived as an interplay between them and their environment. The current state influences which genes are expressed, and gene expression depends on the animal's social and physical world. Individuals are then seen as choosing and changing the conditions to which they are exposed. The question, though, is within what limits will the developmental systems, dynamic as they are, produce the same result. A developmental system that had been sufficiently perturbed would be expected to lead to a markedly different outcome.

Sometimes the perturbations produced by the new set of conditions may be such that developmental processes generate maladaptive outcomes — such as flippers instead of arms when the human embryo had been exposed to thalidomide. If conditions are changed enough, developmental stability is no longer maintained. The biological equivalent of an earthquake occurs and the appearance of organisms may suddenly change. If such a change occurs early in development, the effects may ramify and generate a radically different outcome. Even spontaneously expressed behaviour is only buffered from environmental conditions within certain limits. Changes in conditions may kill the animal, but they can also open up important new avenues for subsequent evolutionary change. That is why knowledge of development impinges on studies of evolution and why genetic determinism has stultified thought about the nature of evolution.

Studies of development and evolution are logically distinct. Knowledge of how a particular automobile has been assembled does not tell us anything about the evolution of automobile design and the same is true of living organisms. The outcome of evolutionary processes is expressed in an individual's development. Furthermore, Darwinian evolution acts on the outcome of developmental processes.[1] The

---

1   The mantra among most evolutionary biologists used to be that evolution involves changes in the frequencies of genes. By contrast Waddington argued that Darwinian evolution acts on the outcomes of development (Waddington, C.H. (1975), *The Evolution of an Evolutionist*. Ithaca, NY: Cornell University Press). In other words evolution involves changes in phenotypes.

growing awareness of the emergent properties of developing systems does, therefore, have implications for evolutionary biologists who traditionally have entertained rather simple notions of what happens in development. Considering all the factors that are involved in development, including the genes, emphasises the benefits of adopting a systems approach.

The likelihood that one group of factors is exclusively important in development and evolution would seem odd to gardeners who have a good understanding of how the character of the soil, the presence of symbiotic activity of fungae next to the roots, the amount of fertilizer supplied, and so forth, all affect how well the plant thrives. Similarly sensitive dog and cat breeders, irrespective of the breed, know how important is the early age of socialisation to the pet's friendliness to humans. People who cook for themselves, know how important is the choice of ingredients that will work well together and on the length and temperature of cooking.

Examples of condition-dependent development do not pose any problems for evolutionary theory, even though they should give pause to those who search for universals within a given species. From a Darwinian standpoint, the development of characteristics that are appropriate to the circumstances in which the individual finds itself makes a great deal of sense.

The many examples of conditional responses to the environment illustrate an important aspect of development that has intriguing implications for humans. Do people have the capacity to live alternative lives? Individual humans are bathed in the values of their own particular culture and become committed by their early experience to behaving in one of many possible ways. Differences in early linguistic experience, for example, have obvious and long-lasting effects. In general, individual humans imbibe the particular characteristics of their culture by learning from older people even if unwittingly.

When environmental conditions influence a particular developmental route in animals, the mechanisms involved are likely to be different; learning may not enter into the picture at all. Even so, is it possible that some aspects of human development are triggered by the environment? Was each individual conceived with the capacity to develop along a number of different tracks each of which is adapted to circumstances in which the individual may find itself? People who grow up in impoverished conditions tend to have a smaller body size, a lower metabolic rate and a reduced level of behavioural activity. These responses to early deprivation are generally regarded as pathological — just three of the many damaging consequences of poverty. The long-term effects on health of a low birth weight may simply be by-products of the continuing social and economic conditions that stunted growth in the first place. Ignorance and shortage of money make the prevention and treatment of disease more difficult; overcrowding, bad working conditions and poverty produce psychological stress and increase the risk of infection. People with little money have poorer diets, and adverse social or physical factors that foster depression and hopelessness increase the risks of disease. In industrialised nations the poor and the unemployed have more illnesses and die sooner than the affluent.

Despite all the well-known effects of poverty, in less extreme conditions they could also be viewed as part of a package of characteristics that are appropriate to the conditions in which the individual grows up — in other words, adaptations to an environment that is chronically short of food, rather than merely the pathological by-products of a bad diet. Having a lower metabolic rate, reduced activity and a smaller body all help to reduce energy expenditure, which can be crucial when food is usually in short supply. To put it simply, they might be adaptations to make the best of a bad job.[2]

If this idea is correct, what about those individuals that are born as big babies who end up in an impoverished environment? The evidence is much weaker, but in general it supports the view that people born in affluent conditions are at greater risk during periods of prolonged famine than those who experienced lower levels of nutrition during prenatal development. Children born to affluent parents are more likely to suffer adverse effects if they are starved in adulthood. In concentration camps and the worst prisoner-of-war camps, anecdotal evidence suggests that the physically large individuals died first while at least some of the small individuals survived. In a famine-exposed Ethiopian population, high birth weight was associated with a nine-fold greater

---

2   Gluckman, P.D., Hanson, M.A. & Buklijas, T. (2010), A conceptual framework for the developmental origins of health and disease. *J. Devel. Origins Health Disease* 1.01, 6–18, https://doi.org/10.1017/s2040174409990171

risk of rickets.³ Rickets in women severely affects their subsequent reproductive success. It seems likely therefore that humans like so many other animals do respond in ways that are usually appropriate to the conditions in which they started life. These ways can be distinctly different from each other.

A series of studies that assessed people across their entire lifespan from birth to death has lent strength to the suggestion that human development involves environmental cues that prepare the individual for a particular sort of environment. Those people who had had the lowest body-weights at birth and at one year of age were most likely to die from cardiovascular disease later in life. They were also more likely to suffer from diseases such as diabetes and stroke in adulthood.⁴ How do these associations these connections make sense in adaptive terms? Could it be that, in bad conditions, the pregnant woman unwittingly signals to her unborn baby that the environment her child is about to enter is likely to be harsh? Such a weather forecast from the mother's body could result in her baby being born with adaptations, such as a small body and a modified metabolism, that help it to cope with a shortage of food. If instead the baby finds itself growing up in an affluent industrialised society, it is poorly adapted.

Charles Darwin's great theory of the evolution of adaptations found in nature remains as important as ever. Even so, the picture of the external hand of natural selection doing all the work is so compelling that it is easy to regard organisms as if they were entirely passive in the evolutionary process. Of course no biologist would deny that organisms, and animals especially, are active. Even so the notion of 'selection pressure' does subtly downplay the role of organisms in the processes of change. When behavioural and developmental issues are joined together with questions about evolution, it becomes easier to perceive how an organism's behaviour can initiate and direct lines of evolution. Developmental processes do not merely act as constraints, they can make certain types of evolutionary change more likely. The

---

3   Chali, D., Enquselassie F. & Gesese M. (1998), A case-control study on determinants of rickets. *Ethiop. Med. J.* 36.4, 227–234

4   Bateson, P., Gluckman, P. & Hanson, M., The biology of developmental plasticity and the Predictive Adaptive Response hypothesis. *J. Physiol.* 592.11, 2357–2368, https://doi.org/10.1113/jphysiol.2014.271460

explosion in the study of epigenetics (see Chapter 8) has suggested some ways in which a link between development and evolution might have occurred.

Proposals about the active involvement of animals in evolution emphasise how their characteristics develop. By contrast, a certain style of evolutionary theory has placed all the emphasis on changes in gene frequencies in the course of evolution, thereby removing the organism from consideration. The justification for this type of theory has been that genes generally survive generation after generation, whereas individual organisms never do. The consequence of such an approach is that, when the effect of a gene on an organism is considered, the gene alone is supposed to determine the outcome.

The advance of epigenetics has awakened interest in the links between development and evolution. The transmitted epigenetic markers across generations could have facilitated genomic change. In most experimental studies, the environmental stimulus producing an epigenetic change is only applied in one generation. Experimentally this may be enough, since research on yeast suggests that an environmental challenge can permanently alter the regulation of genes.[5] In natural conditions, the environmental cues that induce epigenetic change may be recurrent and repeat what has happened in previous generations. This recurring effect might have stabilised the developed characteristics until genomic reorganisation had occurred. The induced epigenetic changes that mediate adaptive plasticity might then have biased the sites of subsequent mutation. Variation at these sites may throw up developed characteristics, some of which are adaptive and subject to Darwinian evolution. This is one way in which plasticity leads to evolutionary change.

Behaviour and evolution

An animal's behaviour is likely to have affected the course of evolution of its descendants in at least four ways. First, animals make active

---

5 Braun, E. & David, L. (2011), The role of cellular plasticity in the evolution of regulatory novelty. In: Gissis, S.B. & Jablonka, E. (eds.), *Transformations of Lamarckism: From Subtle Fluids to Molecular Biology*. Cambridge, MA: MIT Press, pp. 181–191.

choices, and the consequences of their choices are often important. Second, animals change the conditions in which they live by altering the physical or the social environment. Third, active animals often expose themselves to new conditions that reveal variability, with some variants more likely to survive than others. Finally, organisms are adaptable and are able to modify their behaviour in response to novel conditions and thereby make further genetic change possible.[6]

The role of choice in evolution was clearly recognised by Darwin in his principle of sexual selection. He suggested that members of one sex choose to mate with individuals possessing a prominent feature. One example, the tail of the male peacock, has already been mentioned. Other examples are given in Chapter 9. Mate choice sets up a feedback process so that the act of choice affects the evolution of a characteristic in the chosen individual which subsequently then affects what is chosen in descendants. The postulated process could lead to an evolutionary instability with a runaway character.

The crucial agents necessary for this evolutionary process of adaptation driven by choice will generally be elements of the genome, but this is not always the case. In the past the fly *Rhagoletis pomonella* typically laid its eggs on the fruits of hawthorns.[7] Around one hundred and fifty years ago some flies laid their eggs on apples. Their offspring preferred to lay their eggs on apples and the fly has since become a serious pest in USA orchards. The offspring retain through pupation a 'memory' of what they have eaten, and when the new generation of adult flies have mated they lay eggs on the particular species of plant they had eaten before metamorphosis. In this case the variation lies in the behaviour of the adult female flies choosing apples on which to lay their eggs, and onward transmission to the next generation is achieved by an imprinting-like mechanism.

Darwinian evolution operates on characteristics that have developed within a particular set of conditions, many of which are environmental. Apparent design is produced, even when it is at the end of the long and

---

6 The importance of the active role of behaviour in evolution is discussed in Bateson, P. (2013), New thinking about biological evolution. *Biol. J. Linn. Soc.* 112.2, 268–275, https://doi.org/10.1111/bij.12125

7 Michel, A.P. (2010), Widespread genomic divergence during sympatric speciation. *PNAS* 107.21, 9724–9729, https://doi.org/10.1073/pnas.1000939107

complicated process of development involving many different factors. The environment does not cease to be important for evolution just because it remains constant. Change the environment and the outcome of an individual's development may be different. If an individual does not inherit its parents' environment along with their genes and other transmittable factors, it may not be well adapted to the conditions in which it now finds itself. But the altered environmental conditions may throw up variation that was previously hidden, and from that may spring new lines of evolution.

Changes in environmental conditions might, for instance, be imposed by a catastrophe resulting from Earth's collision with a comet or asteroid, or by climate change produced by glaciation and the impact of human activities on the planet. Less dramatically, a change in the environment of a given animal might be brought about because it can move, or in the case of many plants, because their seeds are dispersed. Although the migration of animals can be highly adaptive, the possibility of movement into a novel environment raises a key conceptual point in understanding how developmental plasticity and behaviour can drive evolutionary change.

The organism's contribution towards creating an environment to which it is best suited (see Chapter 9) should give pause if evolution is considered purely in terms of selection by external forces. By leaving an impact on their physical and social environment, organisms may affect the evolution of their own descendants, quite apart from changing the conditions in which they live themselves. Some of the impact is subtle, such as when a plant sheds its leaves which fall to the ground and changes the characteristics of the soil in which its own roots and those of its descendants grow. Some of the impact is conspicuous such as when beavers dam a river, flood a valley and create a private lake for themselves. It has been suggested that the aquatic environment created by the beavers led them to evolve adaptations such as webbed feet that facilitated swimming. The hypothesis is plausible because none of the beaver's nearest relatives, the true gophers and kangaroo rats, have webbed feet. These ideas about the impact organisms on their environment have been developed extensively.

The effect of behavioural control on evolutionary change could be especially great when the social environment is a major component of the challenges faced by animals. The result would be that individuals evolve to understand and predict what other members of their social group are about to do. They become better able to compete with others that do not have this ability. If individuals compete with each other within a social group and the result of the competition depends in part on each individual's capacity to predict what the other will do, the evolutionary outcome might easily acquire a run-away property with the intellectually most advanced individuals driving others in the course of evolution to behave in the same way. Such an explanation, which has been developed eloquently by Nick Humphrey, would fit in with the increase in cranial capacity of humans, assuming that cranial capacity and intellectual ability are correlated.[8]

Nick Humphrey. Photo by LittleHow (2010), Wikipedia, https://commons.wikimedia.org/wiki/File:Nick_Humphrey.jpg, CC-BY SA.

Active control and manipulation of the environment occurs in play. Extended parental care found in birds and mammals may have provided the lift-off for the evolution of increasingly elaborate play with different beneficial outcomes. The surplus energy available to the young might have created optimal conditions for the evolution of the initial appearance of play behaviour. As discussed in Chapter 9 active engagement with the environment has great benefits, because the world is examined from different angles. Such engagement helps to construct a working knowledge of the environment: recognition of objects, understanding what leads to what, discovering that things are found when stones are turned over and the world is rearranged, learning what can and cannot be done with others. All these discoveries are real benefits for the individual, enhancing neural processing, physical fitness, behavioural coordination, and behavioural flexibility.

Those individuals that play more have a putative advantage over the others. In effect they go through a period of training that perfects

---

8   Humphrey, N. (1986), *The Inner Eye*. London: Faber & Faber.

the behaviour they will need when adult. Those of the non-playful individuals' offspring that behave like the playful individuals are more likely to survive and, by degrees, the whole population plays more except when play has over-riding costs. If some individuals are able to profit in ways other than merely improving their motor skills during play, the evolutionary movement towards greater complexity will continue. The more playful individuals might, for example, become more aware of environmental contingencies than others and gain advantage by doing so. Once again this drives the evolution of the same abilities in the rest of the population.

The next step in evolution could, among other things, have led to the ability to generate novel behaviour that would have provided the basis for creativity and innovation. Creative people perceive new relations between thoughts, or things, or forms of expression. They are able to combine them into new forms, connecting the seemingly unconnected. Where does such an evolutionary process stop? Presumably the costs of evolving new forms of behaviour or the sheer difficulties of doing so become limiting.

The effects of a new set of conditions lead either to immediate death or to an appropriate response to the challenge. Initially the response is not inherited, and differential survival of different genotypes may arise from subsequent differences in the ease with which the new character is expressed spontaneously. An unstable evolutionary process could be generated by spontaneous alterations in the genome, but the likelihood of this happening diminishes with the number of components in a response necessary to produce an overall change. The adaptability of the individuals allows by contrast the evolutionary process to occur piecemeal.

Adaptability driving evolution could start operating when a group of organisms respond appropriately to a change in environmental conditions. The modification of form or behaviour occurs generation after generation under the changed conditions, but the modification will not be inherited. Any genetic variation in the ease of expression of the modified characteristic is liable to favour those individuals that express it most readily and with least cost. Consequently, an inherited predisposition to express the modification will tend to evolve. The longer the evolutionary process continues, the more frequent will be such a predisposition. The process starts through learning or some other form of plastic modification within individuals, but this paves the way for a longer-term change in the genome.

In principle, then, behaviour patterns that were initially acquired through the animal's adaptability could be expressed spontaneously, without employing such plasticity, in subsequent generations. It might be argued that spontaneously expressing a behaviour pattern that had been learned in previous generations could be costly if it means that the animal loses its ability to learn. The argument is not cogent when applied to big-brained animals like birds and mammals with multiple parallel pathways for learning. In these animals, the loss of capacity to learn in one way has no effect on the capacity to learn in other ways.

Adaptability can accelerate the rate at which challenges set by the environment can be met. The effect of plasticity on evolution may have become increasingly powerful as animals, in particular, became more complex. Elements could be recombined in different ways to perform different functions. This evolutionary process could lead to the establishment of increasingly elaborate organization and patterns of behaviour. When such complexity entails a greater ability to discriminate between different features of the environment or a greater ability to manipulate the environment, the organism will benefit and become more likely to survive and reproduce in the face of multiple challenges during its lifetime. A new adaptation would emerge in evolution when the accumulated effects of genomic reorganization altered the organism's characteristics. Although these effects are specific to the new function, existing parts of the body's characteristics could also be recruited for this function. Plasticity would promote much more rapid genetic evolution of complex sets of adaptive systems than could be accomplished by changes in the genome. This occurs as previously plastic elements are replaced by inherited elements and the evolving organism is able to replace through its plasticity missing elements in subsequent systems. The exposure to novel environments would be likely to lead to the subsequent evolution by means of classical Darwinian processes of morphological, physiological and biochemical adaptations to those niches.

One case of what can happen when an animal is adaptable has been provided by the three-spine stickleback after moving from a marine to a freshwater environment and then occupying the deep water of lakes or shallow fresh water.[9] It was able to adapt and then developed characteristics that distinguished it from its marine ancestors and which were specialized for the environment into which had moved. Shallow water males have striking red bellies involved in courtship whereas those in deep water, which is dark, do not.

Adaptability to new conditions may be physiological, such as coping with high altitudes by enhancing the oxygen carrying capacity of the blood. Humans living at low altitudes can usually cope when mountaineering, but over many generations this adaptability was followed by inherited genomic change which may take different forms. In the course of evolution people living in the Andes have developed a different response from those living in the Himalayas.[10]

An important empirical demonstration of adaptability driving evolutionary change is that of the house finch. This species is endemic in the western parts of the USA. Some individuals were collected and taken east to New York but were quickly released when the collector realized that he might be prosecuted. The birds adapted and spread north to Canada.

The same species has spontaneously moved north into Montana where it has been intensively studied. After a period involving a great deal of plasticity in a new environment, the house finch populations spontaneously expressed the physiological characteristics that best fitted them to their new habitats without the need for developmental plasticity. Initially the adaptive onset of the time of incubation that occurred in colder climates was affected by the new ambient temperature, but as evolution occurred in the population, these behavioural and physiological

The adaptable house finch. Photo by Thomas Quine (2007), Wikipedia, https://commons.wikimedia.org/wiki/File:Male_House_Finch_profile_(23910087075).jpg. CC BY 2.0

---

9 Foster, S.A. et al. (2015), Evolutionary influences of plastic behavioral responses upon environmental challenges in an adaptive radiation. *Integr. Comp. Biol.* 55.3, 406–417, https://doi.org/10.1093/icb/icv083

10 Beall, C.M. (2007), Two routes to functional adaptation: Tibetan and Andean high-altitude natives. *PNAS* 104.1, 8655–8660, https://doi.org/10.1073/pnas.0701985104

effects were no longer dependent on external cues for their expression. After using their adaptability to respond to the new environmental conditions, the house finch populations spontaneously expressed the characteristics that best fitted them to their new habitats.[11]

Conclusions

Adaptability probably appeared at an early stage in biological evolution. Its role in promoting evolutionary change has not been much investigated. Even so it is plausible that, like the other ways in which an organism's activities can affect its descendants, adaptability has been important. Adaptability can accelerate the rate at which challenges set by the environment can be met. The effect of plasticity on evolution may have become increasingly powerful as animals, in particular, became more complex. Elements could be recombined in different ways to perform different functions. This evolutionary process could lead to the establishment of increasingly elaborate organization and patterns of behaviour.

Darwin's famous metaphor of natural selection is deeply embedded in the modern language of biologists. Natural selection is treated as an agent in much the same way as humans are agents in artificial selection. The strictures on the misuse of the selection metaphor in evolutionary biology will not change many minds since it is not easy to give up the habits of a lifetime. Hopefully, though, the more adventurous will try replacing 'selection' in their writing with 'Darwinian evolution'. This would gives honour where it is due and encourage the view that behaviour does play an active role in evolution.

With the growing acceptance that organisms are not passive in relation to their role in the evolution of their descendants, focus on the adaptability driver helps to bring together studies of development with those of evolution, a principal aim of this book. By doing so, the systems approach provides satisfying explanations for the appearance of design in so much behaviour.

---

11 Badyaev, A.V. (2009), Evolutionary significance of phenotypic accommodation in novel environments: an empirical test of the Baldwin effect. *Phil. Trans. Roy. Soc.* B, 364.1520, 1125–1141, https://doi.org/10.1098/rstb.2008.0285

11. Concluding Remarks

Two themes have run through this book about the development and evolution of behaviour. The first has been about the adaptive processes that give rise to the appearance of design in nature. The second has been about systems, the active role of the organism and the different factors that influence development and evolution. These themes are relevant to human development and some chapters are almost exclusively devoted to human examples.

Not all behaviour is adaptive in the present. In humans the dietary preferences that were adaptive in the past, such as those for salt and sugar, can seriously disrupt health when these substances are readily available.[1] Gambling, which sometimes ruins lives, can seem irrational but makes sense in a world in which the delivery of rewards is rarely random. If you have done something that produced a win, it is usually beneficial to repeat what you did — except when you get into a casino. Similarly, the tendency of parents to protect their children from all contact with unknown people after hearing from the media of a child murder would have been beneficial in a small community where such news might represent real danger. In the modern context, such risk-averse behaviour in a society in which the incidence of child murder has remained constant for decades merely impoverishes the child's development. Even though they were once adaptive, the emotional responses of parents can now have adverse effects on their children's lives.[2]

---

[1] Narvaez, D., Valentino, K., Fuentes, A., McKenna, J., & Gray, P. (2014), *Ancestral Landscapes in Human Evolution: Culture, Childrearing and Social Wellbeing*. New York: Oxford University Press.

[2] The origins of violence in human society involve many inherited dispositions or adverse experiences in early life. The consequences of these developmental

The appearance of design can generate misconceptions. Even so, when a process like behavioural imprinting is examined, it is reasonable to suppose that it plays an important role in the development and survival of each individual bird. Understanding the rules that underlie the development of the individual and the reciprocity between those rules and the individual's experience is important in making sense of the complexity of development.

The young organism has to deal with many challenges that meet it as it develops. Its ecology may be very different from that of the adult, in which case it may have special adaptations to deal with those conditions. Like a caterpillar metamorphosing into a butterfly, a human child has adaptations to deal with each stage of its life cycle. The prevalence of play in the young is an example.

Despite the changes in the individual's repertoire of behaviour as it grows up, early experience can have long lasting effects on its preferences and habits when it finally matures. These aspects of its behaviour are often stable, but in stressful conditions they may change when the stress is accompanied by new forms of experience. The change can be adaptive since it can enable the individual to cope with a world that may be very different from the one in which it grew up.

In mammals, parent and offspring are often thought to be in conflict. On this view, the communication between them takes the form of mutual manipulation. The offspring seeks to gain maximum advantage from its parent and the parent seeks to defend its long-term reproductive interests. Against this view, communication is often such that both the sender and receiver of a signal or cue may benefit by both parties treating it as useful to themselves. In the case of parent-offspring relations, parents do well to take into account the condition of their offspring and the offspring must likewise pay attention to the condition

---

abnormalities are expressed as harmful or psychopathic behaviour already, before the age of three, and may persist throughout life. However, in a humane society much can be done to help such troubled people by identifying them early on, giving their parents extra support, treating them with sensitivity and not punishing them for bad behaviour. They need not be treated as irredeemable and effectively given a life sentence. An excellent discussion of the origins of anti-social violence is given by Adrian Raine (2014), *The Anatomy of Violence: The Biological Roots of Crime*. London: Penguin Books.

of their parent. In other words, their behaviour is adaptive for both parties.

Many animals choose their mates carefully. This is especially true in birds and many mammals. Inbreeding has costs but so too does excessive outbreeding. The way in which an optimal balance is achieved is in part through experience with close kin in early life leading to a preference for a mate who is a bit different but not too different from familiar kin. An important issue is whether the avoidance of incest found in most human societies serves the same function as the avoidance of inbreeding. A common function is questionable and the taboos or more likely to be an expression of conformism directed at individuals doing what most people would not do.

With the great successes of molecular biology, attention has been focused on the role of genes in development. Genes are unquestionably needed for the inheritance of much behaviour. The importance of genes, however they are defined, does not mean that a simple link can be found between genes and behaviour. The links are usually complex and metaphors such as genes providing a blueprint for behaviour are misleading. The benefits of the selfish gene approach in understanding the complexities of evolution do not imply that genes program the development of an individual. Understanding development requires a systems approach that takes into account all the genes and environmental inputs that are effective.

Development and evolution are usually regarded as separate domains of inquiry. Even so an organism's adaptability provides a useful link between these domains. It offers understanding of the relationship between what an individual does and how its activities might influence the genomes of its descendants. Many theoretical arguments have been used to explain how this might happen and some empirical evidence supporting these arguments is becoming available.

The notion of genetic determination, which is so firmly embedded in evolutionary theory, has seriously interfered with attempts to understand the dynamics of behavioural development and its role in evolution. If anything has been learned in recent years, it is that what an individual animal does in its life is conditional and depends on the reciprocal character of the transactions with the world about it. This knowledge also points to ways in which an animal's own behaviour can provide the variation that influences the subsequent course of evolution. When developmental issues are joined with questions about evolution, it becomes easier to perceive how an organism's behaviour can initiate and shape evolutionary change.

These changes in biological thinking affect the relations between the natural and the social sciences. The biggest block to bringing the biological and social sciences together was the presumption that Darwinian evolution implied genetic determinism. This block has now been removed by advances in biological thought (see Chapter 8). Behavioural biologists have sometimes misleadingly applied terms such as 'greed', 'spiteful', 'rape', 'marriage' and 'incest' to animals. This may have been done to lighten the normally dull language of scientific discourse. However, these terms have an established usage in describing human emotions, and in describing human institutions with all their associated rights, individual responsibilities and culturally transmitted rules on what people may and may not do. Problems of communication between disciplines have been compounded when, having found some descriptive similarities between animals and humans, and having investigated the animal cases, biologists or their popularisers have used the animal findings to 'explain' human behaviour. Such arguments rely on a succession of slippages in meaning and are usually unconscious, but they provoke hostility in those people in the social sciences and humanities who feel threatened by an apparent take-over bid on the part of biologists. An example of how the effects of early experience promoted by biologists have been misconstrued by the social scientists is described in Chapter 5.

The conflicts of motivation evident in studies of animal behaviour bear on important issues to do with human behaviour. In many social contexts a person might weigh up consciously or unconsciously the benefits to themselves of behaving in a particular way. The benefits might include avoiding disapproval or punishment by other people. However, all these dispositions can conflict with powerful impulses to act in ways that benefit the social group, the tribe, or some larger assemblage without any direct benefits to the individual. They may cooperate with individuals they regard as belonging to their own social group and express fear or hate of those individuals they regard as being different or foreign. They may be influenced by their desire for leadership and strongly disapprove of anybody who does not conform with their views. None of these impulses are invariant. They can be, and often are, changed by experience. Conformism and the expression of fear and hate can be inhibited, or modified either by social norms or by becoming aware of the damaging consequences of allowing full reign to such impulses.

By bringing together evidence from different areas of knowledge, more powerful theoretical perspectives can be formulated. Their impact is not only on scientific approaches to the systems of development and evolution, but also on how humans change institutional rules that have become dysfunctional or design public health measures when mismatches occur between themselves and their environments. It affects how humans think about themselves and their own capacity for change. The biological approach to human psychology does not imply that individuals do not have free choices. Through their decisions individuals clearly do make a big difference to what happens in their lives. They may be sometimes surprised by the consequences of their own actions. Even so, they are able to anticipate the consequences of various courses of action and choose between them on the basis of their likely costs and benefits. Planning before doing is clearly of great advantage. The evidence stares us in the face. People do make well-considered decisions and they benefit from doing so.

Index

accommodation 99, 113
active role 5, 32, 80, 107, 113, 115
adaptability 5, 14, 17, 35, 98, 100, 110, 111, 112–113, 117
adaptations 12, 16, 98, 107, 111–112
aggression 6, 61–62
Albon, S.D. 61
allele 89
Aronson, L. 28
attachment 3, 19, 22, 23, 24. *See also* bonding
attention 4, 6, 22–23, 36, 64–65, 73, 116–117

bacteria 10, 81, 89
Badyaev, A.V. 113
Baroncelli, L. 54
Beall, C.M. 112
beavers 108
behavioural imprinting 6, 116
behavioural studies 6, 19, 30, 34, 57, 62, 69, 96, 105–106, 118
  determinism 47
behaviour patterns 1, 10, 17, 27–28, 31, 33–34, 41–44, 49, 51, 54, 57, 59, 79–81, 89, 91, 93–94, 111–112, 113
  environmental 79, 84
  genetic 1, 27, 79–80, 84, 85, 117
  innate 1, 27–28
  learned 1, 14, 27, 91, 111
  pattern recognition 14–15, 24
birds 2, 19–20, 23–25, 32–33, 71, 87, 116

Bittles, A.H. 67
blastula 39
blueprint 4, 80, 117
Blumberg, M.S. 5
bonding 52–53. *See also* attachment; parent-offspring relationships
  animal 4, 94, 117
Book of Common Prayer 74
Bowlby, John 21
brain 6, 12, 15, 21, 36, 42, 99
brains 49, 50, 53, 54
brainwashing 50–52
Braun, E. 106
Buklijas, T. 104
butterfly 4, 40, 116

Capra, F. 3, 29
Carey, N. 85
caterpillar 4, 40, 116
cats 17, 53, 54, 63, 94
Chali, D. 105
cheetah 91, 94
chess 13–15, 101
  Deep Blue 13–14
  Deeper Blue 14–15
chimpanzee 71, 84, 95
choice 12, 20, 33, 69, 75, 91, 103, 107
chromosomes 78, 86
Clarke, A.D.B. 47
Clarke, A.M. 47
cleaner fish 58
Clutton-Brock, T.H. 61

communication 4, 57–58, 61, 116, 118
competition 14–16, 57, 92, 94, 109
computer 2, 13–15. *See also* chess
  IBM 13–15
conditional 35, 64–65, 79, 100, 103, 117
conflict 4, 15, 59–62, 116, 118
conformity 74–75
convergers 96
cooperation 14, 57–58
coping 99, 112
courtship 68, 112
cousins 67, 70, 71, 75
cow 83–85
creativity 5, 6, 13, 93, 96–97, 110
Cresswell, W. 91
cuckoo 57

Darwin, Charles 3, 10, 59, 88, 91, 105, 107, 113
Darwinian evolution 10, 17, 35, 57, 69, 73, 75, 87, 97–98, 102, 106, 107, 113, 118
David, L. 106
Dawkins, Richard 57, 87
descendants 5, 106–107, 113, 117
design 3, 9–12, 17–18, 45, 55, 91, 98, 102, 107, 113, 115–116, 119
determinism 47, 102
development 1, 3–6, 11–14, 16, 19–22, 24–25, 27–38, 39–46, 47–49, 55–56, 57, 62, 63, 65–66, 77, 79–81, 83–87, 89, 91, 94, 97–100, 101–106, 108, 113, 115–118
  cake model 79
  discontinuous 39
  sensitive periods in 20, 21, 47–49, 55
developmental studies 27
  controversies 27
differentiation 86
discontinuities 4, 39–41, 44–45, 63–64
divergers 96
DNA 77–78, 80–81, 85–86
  'junk' DNA 81

dogs 52, 99, 103
Durham, W.H. 67, 73–74, 79

ecology 3, 16, 17, 39, 45, 116
Edwards, Jonathan 50
embryo 1, 39, 85, 102
emperor penguin 16
Enquselassie, F. 105
environment 1, 5, 12, 20, 27–28, 32, 35–37, 47, 55, 64–65, 69–70, 74, 77, 80, 82–86, 92, 97–98, 100, 101–105, 107–109, 111–113
epigenetic 28–29, 31, 50, 85–86, 97, 106
equifinality 29–31
ethology 2, 5–6, 27, 39
evolution 3, 5–6, 10–12, 17, 21, 26, 27, 31, 35, 57, 61, 69, 73, 74, 75, 77, 77–78, 78, 81, 85, 87–89, 97–98, 102–103, 105–113, 115, 117–118
  biscuits 88
  evolvability 88

Fanshawe, J.H. 91
feedback 2, 29–32, 35, 107
females 52, 53, 71, 95–96, 107
Ferguson-Smith, A.C. 85
Fitzgibbon, C.D. 91
food 16, 17, 29–30, 40, 41, 43, 49, 54, 58, 59, 61–62, 64, 98, 104–105. *See also* weaning
forecasting 88, 105
Foster, S.A. 112
Freud, Sigmund 48
Fuentes, A. 115
fur seals 63

gambling 115
  poker 60
gastrula 39
gene 3–5, 17, 27–28, 67–68, 71, 73, 77–89, 97–98, 101–103, 106, 108, 117
genetic 1, 27, 67–70, 75, 78–80, 82, 83, 84, 85, 86, 88, 89, 91, 99, 102, 107, 110–111, 117–118

genetic-environment dichotomy 1
genomes 5, 78–79, 84, 97–98, 107, 110–111, 117
Gesese, M. 105
Gissis, S.B. 85, 106
Gluckman, P.D. 104–105
Gomendio, M. 62, 94
Gottlieb, G. 20
Grafen, A. 60
grand-offspring 69, 75–76
Gray, R.D. 2, 115
Greenwood, P.J. 70
Griffiths, P.E. 2, 78

habituation 6, 73
Hanson, M.A. 104–105
height 42, 69, 82–83
Helgasson, A. 70
heterarchy 16
Heyes, C. 24
hierarchy 16
  rank 40–44
Hinde, Robert A. 6, 19, 34
Holmes, J. 22
honey bees
  waggle dance 58
Horn, Gabriel 3, 6, 20
house finch 112, 113
Hoyle, G. 31
Huber, L. 24
Human Genome Project 78
Humphrey, Nick 109
hybrid vigour 69

identical twins 84, 85
imprinting 19–26, 32, 73, 107. *See also* behavioural imprinting
  filial imprinting 24, 25
  in the laboratory 20, 22, 23
  in the wild 22
  sexual imprinting 24–25, 69
inbreeding 4, 67–68, 70, 71, 73–75, 117
incest 4, 67, 73–76, 117–118
  incest taboo 73–74, 75

information 16, 25, 32, 35, 57–60, 65, 66, 79, 86, 97
inherited 1, 10–11, 68, 77, 82, 86, 110–112
innate 1, 27–28
innovation 97, 110
instinct 27
interplay 2–3, 65, 98, 101–102
IQ 42–43
Israeli kibbutzniks 72

Jablonka, E. 85, 106
Jackson, P.S. 24
Japanese quail 71
Johnson, M.H. 20
jungle fowl 20

Kacelnik, A. 20
Kahlenberg, S.M. 95
Kasamutsu, T. 53
Kasparov, Gary 13–14
Keller, E.F. 78
kittens 40, 41, 43, 63, 93
Klopfer, P.H. 39
Knudsen, E.I. 21
Korean War, the 50
Krebs, J.R. 57

Laland, K. 11
learning 6, 11, 17–18, 19–21, 23–26, 28, 35, 71–73, 97–98, 101, 103, 109–111
Lehrman, D.S. 28
Lewontin, R. 79
local optima 95
Luisi, P.L. 3, 29

Maccoby, E.E. 72
major marriage 71, 72
males 16, 60, 71, 91, 95–96, 107, 112
Mameli, M. 28
mammals 4, 10, 11, 25, 62, 65, 67, 70, 71, 81, 93, 109, 111, 116, 117
Marconnato, F. 20
Martinho, A. 20
Martin, Paul 6–7, 13, 40, 93, 97

mating 67–68, 70, 75, 94
McKay, D. 55
McKenna, A. 115
memory 6, 13, 44, 107
Mendelian inheritance 82
metabolic rate 104
metamorphoses 40, 43
metaphors 4, 17, 32, 79, 87–88, 96, 113, 117
mice 81
Michel, A.P. 107
minor marriage 71, 72
Miska, E.A. 85
mobility 91–92
mole 9
Molenaar, P.C.M. 5
mother/child relationship 21–23, 34, 62–66, 89. *See also* parent/offspring relationship
  bonding 21
    in animals 34
    in birds 22, 23
    in humans 21
mountain delphinium 69
Munz, T. 58
Murray, L. 21
mutation 69, 77, 98, 106

Narvaez, D. 115
natural selection 3, 10, 88, 105, 113
nature vs nurture 27, 101
negotiation 60
Nettle, D. 96
neurobiology 53–54
neurons 53–54
niche construction 92
  niches 111
Noble, Denis 77
noradrenaline 53–54
nutrition 63–64, 82, 104

offspring 3–4, 16–17, 25, 33, 40, 57, 61–66, 69–71, 74, 75, 82, 89, 95, 107, 110, 116, 117
Oliverio, A. 47

optimal outbreeding 69–70, 75
organism 3, 12, 18, 35–37, 77, 79–81, 85, 86, 87, 91–92, 99–100, 105, 106, 108, 111, 113, 115–117
outbreeding 4, 67–70, 75–76, 117
Oyama, S. 2, 79

Paley, William 9–10
parent/offspring relationship 3–4, 17, 40, 62–66, 82, 95, 116
parents 1, 3–4, 19, 23–26, 30, 35, 43, 49, 57, 61–62, 65, 69, 75, 82, 89, 97, 100, 104, 108, 115–116. *See also* bonding
  surrogate 23–24
peacock 91, 107
personality 43, 44, 49, 79
Pettigrew, J.D. 53
phobia 51
pig 9, 83
plasticity 1, 34–35, 49–53, 55, 84–85, 99–100, 105–106, 108, 111–114
play 5–6, 24, 27, 49, 62, 72, 75, 80, 93–97, 109, 110, 113, 116
  acquisition of experience 94–95
  and creativity 96, 110
  and parental overprotection 115
  in animals 93–94
  in humans 93–94, 96
  socialising role 93, 94
polypeptides 80–81
predation 70, 91, 93
predisposition 20, 21, 110, 111
preferences 4, 20–21, 22, 30, 49, 55, 72–73, 89, 115–116
Price, M.V. 69
primatology 6
proteins 80
psychotherapy 51
puberty 39, 43

rat 29, 64
reciprocity 3, 27, 36, 37, 116
recognition of kin 24, 25
red deer 60, 61

Reese, E.P. 23
Regolin, L. 20
reproductive success 16, 61, 63, 64, 65, 67, 73, 88, 105
Rhagoletis pomonella 107
rhesus monkeys 34
rickets 105
RNA 78, 80–81, 85–86
robustness 1, 36
Rosenblatt, J.S. 28
Rowell, C.H.F. 87

Sargant, William 51–52
selection pressure 105
sensitive period 20, 21, 48, 55
sexual selection 107
Shapiro, J. 85
sheep 52, 53
Shenk, D. 5
Shields, J. 84
signalling 59–60, 63. *See also* communication
Simpson, M.J.A. 34
social insects 65. *See also* honey bees
Spector, T. 84
Spencer, J.P. 5
spouse 69, 72
squabbling 62, 65
Stevenson Hinde, J. 19
stickleback 111
Stockholm syndrome 52
Stotz, K. 78
Strohl, K.P. 44
Sultan, S. 78
systems 2, 3, 4, 12, 14–16, 21, 26, 28–29, 31–33, 36–37, 41, 44, 54, 70, 80–81, 85, 88, 89, 94, 102–103, 111, 113, 115, 117–118

Taiwan 71, 72
tantrums 62
teeth 9, 58, 68, 99
thalidomide 99, 102
therapy 48, 51, 55
    'flooding' 51
Thomas, A.J. 44
Tinbergen, Niko 6, 11
Tobach, E. 28
transactions 28, 117
trauma 51, 51–53, 52, 53
Trevarthen, C. 21
Trivers, Robert 61
Turner, D.C. 43

United States of America 50, 51, 53, 61, 67

Valentino, K. 115
Vallortigara, G. 20
Vidal, J.-M. 24
visual system 53

Waddington, C.H. 3, 28–29, 31, 50, 85, 102
Waser, N.M. 69
weaning 39, 43, 61–65
weather 17, 105
Westermarck, E. 74, 75
Wilson, E.O. 73
Wolf, A.P. 67, 72, 75
Wrangham, R.W. 95
Wright, S. 69

Yudkin, Michael 7

Zahavi, Amotz 59, 60
Zahavi, Avishag 59
Zappella, M. 47

This book need not end here…

At Open Book Publishers, we are changing the nature of the traditional academic book. The title you have just read will not be left on a library shelf, but will be accessed online by hundreds of readers each month across the globe. OBP publishes only the best academic work: each title passes through a rigorous peer-review process. We make all our books free to read online so that students, researchers and members of the public who can't afford a printed edition will have access to the same ideas.

This book and additional content is available at:
https://www.openbookpublishers.com/product/490

Customize

Personalize your copy of this book or design new books using OBP and third-party material. Take chapters or whole books from our published list and make a special edition, a new anthology or an illuminating coursepack. Each customized edition will be produced as a paperback and a downloadable PDF. Find out more at:

http://www.openbookpublishers.com/section/59/1

Donate

If you enjoyed this book, and feel that research like this should be available to all readers, regardless of their income, please think about donating to us. We do not operate for profit and all donations, as with all other revenue we generate, will be used to finance new Open Access publications.

http://www.openbookpublishers.com/section/13/1/support-us

Like Open Book Publishers

Follow @OpenBookPublish

Read more at the Open Book Publishers BLOG

You may also be interested in…

*Animals and Medicine: The Contribution
of Animal Experiments to the Control of Disease*
By Jack Botting

http://dx.doi.org/10.11647/OBP.0055
http://www.openbookpublishers.com/product/327

The Scientific Revolution Revisited
By Mikuláš Teich

http://dx.doi.org/10.11647/OBP.0054
http://www.openbookpublishers.com/product/334

*Measuring the Master Race:
Physical Anthropology in Norway, 1890-1945*
By Røyne Kyllingstad

http://dx.doi.org/10.11647/OBP.0051
http://www.openbookpublishers.com/product/123

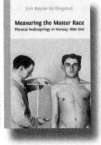

Lightning Source UK Ltd.
Milton Keynes UK
UKOW06f1249080217

293864UK00004B/1/P